Laurier L. Schramm

**Dictionary of Nanotechnology,
Colloid and Interface Science**

Further Reading

L. L. Schramm

Emulsions, Foams, and Suspensions

Fundamentals and Applications

2005
ISBN: 978-3-527-30743-5

T. F. Tadros (Ed.)

Colloids and Interface Science Series

6 Volumes

2007–2009
ISBN: 978-3-527-31461-4

T. F. Tadros (Ed.)

Topics in Colloid and Interface Science

2009–2010
Volume 1: ISBN: 978-3-527-31990-9
Volume 2: ISBN: 978-3-527-31991-6

C. N. R. Rao, A. Müller, A. K. Cheetham (Eds.)

Nanomaterials Chemistry

Recent Developments and New Directions

2007
ISBN: 978-3-527-31664-9

E. Ruiz-Hitzky, K. Ariga, Y. M. Lvov (Eds.)

Bio-inorganic Hybrid Nanomaterials

Strategies, Syntheses, Characterization, and Applications

2008
ISBN: 978-3-527-31718-9

Laurier L. Schramm

Dictionary of Nanotechnology, Colloid and Interface Science

WILEY-VCH

WILEY-VCH Verlag GmbH & Co. KGaA

The Author

Dr. Laurier L. Schramm
511 Braeside View
Saskatoon, SK, S7V 1A6
Canada

All books published by Wiley-VCH are carefully produced. Nevertheless, authors, editors, and publisher do not warrant the information contained in these books, including this book, to be free of errors. Readers are advised to keep in mind that statements, data, illustrations, procedural details or other items may inadvertently be inaccurate.

Library of Congress Card No.: applied for

British Library Cataloguing-in-Publication Data
A catalogue record for this book is available from the British Library.

Bibliographic information published by the Deutsche Nationalbibliothek
The Deutsche Nationalbibliothek lists this publication in the Deutsche Nationalbibliografie; detailed bibliographic data are available on the Internet at http://dnb.d-nb.de.

© 2008 WILEY-VCH Verlag GmbH & Co. KGaA, Weinheim

Printed in the Federal Republic of Germany
Printed on acid-free paper

Typesetting Thomson Digital, Noida, India
Printing betz-druck GmbH, Darmstadt
Binding Litges & Dopf GmbH, Heppenheim

ISBN: 978-3-527-32203-9

Contents

Dictionary of Nanotechnology, Colloid and Interface Science. Laurier L. Schramm
Copyright © 2008 WILEY-VCH Verlag GmbH & Co. KGaA, Weinheim
ISBN: 978-3-527-32203-9

About the Author

Dr. Laurier L. Schramm is President and CEO of the Saskatchewan Research Council (SRC), a leading Canadian provider of applied research, development, design, and technology demonstration. He previously served as Vice President, Energy of the Alberta Research Council, President and CEO of the Petroleum Recovery Institute, and as a senior scientist in industrial process research at Syncrude Canada Ltd. He has served on the Boards of Directors, or their equivalent, of numerous organizations engaged in research, development, and demonstration, and was most recently a member of the National Panel of Experts on Sustainable Energy Science & Technology (Canada).

Dr. Schramm has been a strong proponent and practitioner of university-industry collaboration and is an adjunct professor of chemical and petroleum engineering at the University of Calgary. He has over 30 years of R&D experience in colloid, interface, and petroleum science, and has received major national awards for his research. He holds 17 patents, over 300 other scientific publications and proprietary reports, and has given over 100 national and international scientific presentations. Many of his inventions have been adopted into commercial practice; and his work on the development of oil-tolerant foams for enhanced oil recovery was judged to be a Milestone of Canadian Chemistry in the 20th Century by the Canadian Society for Chemistry. In 2002 he was awarded Canada's Queen Elizabeth II Golden Jubilee Medal for his creative contributions to Canadian petroleum technology. This is his ninth book.

Dictionary of Nanotechnology, Colloid and Interface Science. Laurier L. Schramm
Copyright © 2008 WILEY-VCH Verlag GmbH & Co. KGaA, Weinheim
ISBN: 978-3-527-32203-9

Acknowledgments

This book was made possible through the support of my family, Ann Marie, Katherine, Victoria, and my parents William and Joan, all of whom have provided consistent encouragement and support.

I thank my colleagues who invested considerable time and effort reviewing earlier editions of this book and contributing comments and suggestions. There are too many such colleagues to list here but Elaine Stasiuk, Randy Mikula, Susan Kutay, and the late Karin Mannhardt and Loren Hepler were all especially helpful. I also thank the staff of Wiley & Sons Inc. in the USA and Wiley-VCH in Germany, especially Darla Henderson, Karin Sora, Martin Ottmar, Rainer Münz, and Claudia Grössl.

Because there are so many different, specialized references to aspects of nano-technology and colloid and interface systems, especially in industrial practice, some important terms will inevitably have been missed. I will greatly appreciate it if readers would take the trouble to inform me of any significant errors or omissions.

Saskatoon, SK, Canada   *Laurier L. Schramm*
July 2008

Dictionary of Nanotechnology, Colloid and Interface Science. Laurier L. Schramm
Copyright © 2008 WILEY-VCH Verlag GmbH & Co. KGaA, Weinheim
ISBN: 978-3-527-32203-9

Introduction and Historical Evolution

In the early 1800s Thomas Graham studied the diffusion, osmotic pressure, and dialysis properties of a number of substances, including a variety of solutes dissolved in water (see References [1–3]). He noticed that some substances diffused quite quickly through parchment paper and animal membranes and formed crystals when dried. Other substances diffused only very slowly, if at all, through the parchment or membranes and apparently did not form crystals when dried. Graham proposed that the former group of substances, which included simple salts, be termed "crystalloids", and that the latter group, which included albumen and gums, be termed "colloids". Although colloidal dispersions had certainly been studied long before this time, and the alchemists frequently worked with body fluids, which are colloidal dispersions, Graham is generally regarded as having founded the discipline of colloid science.

The test of crystal formation later turned out to be too restrictive, the distinction of crystalloids versus colloids was dropped, and the noun colloid was eventually replaced by the adjective colloidal, indicating a particular state of dispersed matter: matter for which at least one dimension falls within a specific range of distance values. The second property that distinguishes all colloidal dispersions is the extremely large area of the interface between the two phases compared with the mass of the dispersed phase. Table 1 in Appendix 1 illustrates the wide range of dispersions concerned. It follows that any chemical and physical phenomena that depend on the existence of an interface become very prominent in colloidal dispersions. Interface science thus underlies colloid science.[1] In 1917 Wolfgang Ostwald, another founder of colloid science, wrote:

---

1) Here again we encounter evidence of a living language. Following Graham's identification of a new division of chemistry, colloid chemistry, the realiza-tion of the profound importance of the interface between the phases led subsequent chemists to refer to the discipline as colloid and capillary chemistry (meaning colloid and interface chemistry). In view of the wide interdisciplinary nature, I prefer the term colloid and interface science.

Dictionary of Nanotechnology, Colloid and Interface Science. Laurier L. Schramm
Copyright © 2008 WILEY-VCH Verlag GmbH & Co. KGaA, Weinheim
ISBN: 978-3-527-32203-9

"It is simply a fact that colloids constitute the most universal and the commonest of all the things we know. We need only to look at the sky, at the earth, or at ourselves to discover colloids or substances closely allied to them. We begin the day with a colloid practice – that of washing – and we may end it with one in a drink of colloid coffee or tea" [4].

Now, more than 200 years later, a vast lexicon is associated with the study of colloid and interface science because, in addition to the growth of the fundamental science itself, we recognize a great diversity of occurrences and properties of colloids and interfaces in industry and indeed in everyday life. Many other scientific disciplines become involved in the study and treatment of colloidal systems, each discipline bringing elements of its own special language. The most recent addition is the overlapping field of nanotechnology.

In 1959, physicist Richard Feynman gave the first known lecture on nanotechnology at the annual meeting of the American Physical Society [5], in which he proposed the idea that atomic manipulation could be used to build structures. The term nanotechnology itself appears to have been coined in Japan, in 1974, by Norio Taniguchi to describe processes at the nanometre scale. Significant interest, and work in, the area of nanotechnology grew particularly rapidly following the publication of the book Engines of Creation by Eric Drexler in 1986 [6]. An illustration of the new way of thinking that is represented by nanotechnology is given in this quote from B.C. Crandall:

"We are distinct from all previous generations in that we have *seen* our atoms – with scanning tunneling and atomic force microscopes. But more than simply admiring their regular beauty, we have begun to build minute structures. Each atom is a single brick; their electrons are the mortar. Atoms, the ultimate in material modularity, provide the stuff of this new technology" [7].

Now, just over 20 years later, the nanotechnology field has grown exponentially, leading to a plethora of new terms in the areas of nanoscience and nanotechnology. The "nano" regime, which spans the distance range from 1 to 100 nm, explicitly overlaps heavily with the size-range of colloid science and technology, which spans the range from 1 to 1,000 nm. As well, some authors distinguish between nanotechnology and microtechnology, the latter referring to species and phenomena in the micrometre scale. There has been an explosion of terms with the "nano" prefix and the number of possible "nano" terms is virtually unlimited, especially when material types are included (Table 2 in Appendix 1 provides an illustrative listing and Table 3 shows the prefix nano in relation to other decimal prefixes in science and technology). For example, there are a wide range of types of nanorods, nanotubes, nanowires, nanobelts, and nanoribbons in nanoscale electronic circuit elements alone. Accordingly some choices have had to be made regarding how many "nano" terms to include in this book.

Although it is true that some nanodispersions are simply colloidal dispersions under a new name, many other aspects of nanotechnology are genuinely new and distinct, such as carbon nanotubes and quantum dots. It has been suggested, but not universally adopted, that the term nanotechnology be used to refer to the study of the nanoscale regime, and the term molecular nanotechnology be used to refer to the 'nano approach,' by which is meant the precise, controlled assembly of structures up from the molecular scale that are well-organized. This is in contrast to the classical "top down" approach of making things by cutting, bending, and otherwise shaping structures from larger starting pieces. In the dispersions area an analogy would be the use of colloidal ink dispersions in robocasting to build near-nanometre scale three-dimensional structures, as opposed to the formation of materials by subdividing bulk phases and then kinetically stabilizing their dispersions using emulsifiers and stabilizers.

This book provides brief explanations for the most important terms that may be encountered in a study of the fundamental principles, experimental investigations, and industrial applications of nanotechnology and colloid and interface science. Even this coverage represents only a personal selection of the terms that could have been included were there no constraints on the size of the book.

I have tried to include as many important terms as possible, and cross-references for the more important synonyms and abbreviations are also included. The difficulty of keeping abreast of the colloid and interface science vocabulary, in particular, has been worsened by the tendency for the language itself to change as the science has evolved since the 1800s, just as the meaning of the word colloid has changed. Many older terms that are either no longer in common use, or worse, that now have completely new meanings, are included as an aid to the reader of the older colloid and interface science literature and as a guide to the several meanings that many terms can have. As an emerging field, the meanings of terms in nanoscience and nanotechnology are still somewhat in flux, although some standardization is beginning to occur. As an indication of the continual evolution of the science, technology, and terminology of the "very small," this book also contains a modest number of terms from the emerging field of picotechnology.

Some basic knowledge of underlying fields such as chemistry, physics, geology, and chemical engineering is assumed. Many of the important named colloids and phenomena (such as Pickering emulsions), equations, and constants are included, although again this selection represents only some of the terms that could have been included. Finally, I have also included a selection of brief biographical introductions to more than 85 scientists and engineers whose names are associated with famous named phenomena, equations, and laws in nanotechnology and colloid and interface science. Students first become aware of the people that have laid the foundation for a scientific discipline as they encounter these eponyms. By adopting the "students' view" of famous names in the field, it will be seen that in some cases the scientists are very famous, and biographies are readily found. In other cases the scientists are not as well known, and in some cases their contribution to nanotechnology and colloid and interface science was otherwise slight. For those interested in this feature specifically, I have included an index of famous names in nanotechnology and colloid and interface science (Table 21 in the Appendix) for easy searching.

Specific literature citations are given when the sources for further information are particularly useful, unique, or difficult to find. For terms drawn from fundamental colloid and interface science, much reliance was placed on the recommendations of the IUPAC Commission on Colloid and Surface Chemistry [8]. Numerous other sources have been particularly helpful in colloid and interface science (textbook [9–14]) and its subdisciplines and related, specialized fields [15–31]. I recommend these sources as starting points for further information. Similarly, for terms emerging in nanotechnology, much reliance was placed on the recommendations of ASTM Committee E56 on Nanotechnology [32] and the British Standards Institution Vocabulary on Nanoparticles [33]. Other helpful sources include [34–38]. For the famous names entries, I have drawn on a number of general references [39–44] and have also included numerous specific references for those interested in additional information.

A

AAN	$\rightarrow$ Average Agglomeration Number.
Ablation	The reduction of particles into smaller sizes due to erosion by other particles or the surrounding fluid. May also refer to the size reduction of liquid droplets due to erosion, as in the processing of an oil sand slurry in which the oil (bitumen) is very viscous.
Abrasion	The wearing down of a surface by erosion due to particles in the surrounding fluid.
Absolute Filtration Rating	The diameter of the largest spherical particle that will pass through a filter, under given test conditions, without deformation.
Absolute Viscosity	A term used to indicate viscosity measured by using a standard method, with the results traceable to fundamental units. Absolute viscosities are distinguished from relative measurements made with instruments that measure viscous drag in a fluid without known or uniform applied shear rates. $\rightarrow$ Viscosity, *see* Table 4.
Absorbance	In optics, a characteristic of a substance whose light absorption is being measured. The Beer–Lambert law gives the ratio of transmitted (I) to incident (I_0) light as $\log(I/I_0) = alc$, where a is the absorptivity, l is the optical path length, and c is the concentration of species in the optical path. The logarithmic term is called the "absorbance".
Absorbate	A substance that becomes absorbed into another material, or absorbent. $\rightarrow$ Absorption.
Absorbent	The substrate into which a substance is absorbed. $\rightarrow$ Absorption.
Absorption	The increase in quantity (transfer) of one material into another or of material from one phase into another phase. Absorption may also denote the *process* of material accumulating inside another.

Dictionary of Nanotechnology, Colloid and Interface Science. Laurier L. Schramm
Copyright © 2008 WILEY-VCH Verlag GmbH & Co. KGaA, Weinheim
ISBN: 978-3-527-32203-9

Acacia Gum	→ Gum.
Accumulation Aerosol	An aerosol in which the primary particles or droplets have aggregated and/or coalesced into larger species or aggregates, typically in the size range of 50 to 1,000 nm. → Aerosol and → Nucleation Aerosol.
Acheson, Edward (Goodrich) (1856–1931)	An American electrochemist and inventor known for his work in the electrical and electric lighting fields, and in abrasives. A contemporary of Thomas Edison, with whom he was at times a collaborator or a competitor, Acheson developed conducting carbon for Edison's electric light bulbs, and managed electric generating plants and lamp manufacturing factories in Europe and the United States. Acheson discovered silicon carbide, its practical application as an abrasive, and coined the name Carborundum. Acheson also developed pure graphite and colloidal graphite products and founded several companies for their manufacture. Two of Acheson's colloidal graphite products (suspensions in oil or water) were called Oildag and Aquadag. He was granted 70 patents on devices, techniques, and compositions of matter in the fields of mechanics, electricity, electrochemistry, and colloid chemistry.
Acicular Particle	A long, narrow particle, such as a "needle-shaped" particle. Example: pine needles.
Acid Number	→ Total Acid Number.
ACN	Alkane carbon number, → Equivalent Alkane Carbon Number.
Acoustophoretic Mobility	An analogue of the electrophoretic mobility that can be calculated from either of the electroacoustical methods of electrokinetic sonic amplitude or ultrasound vibration potential. → Electrokinetic Sonic Amplitude, → Ultrasound Vibration Potential, and *see* Reference [45].
Activated Adsorption	Chemisorption, that is, adsorption for which an activation energy barrier must be overcome, as opposed to unactivated adsorption, or physisorption, for which there is no activation energy barrier to be overcome. → Chemisorption, → Physisorption.
Activated Carbon	Carbonaceous material (such as coal) that has been treated, or activated, to increase the internal porosity and surface area. This treatment enhances its sorptive properties. Activated carbon is used for the removal of organic materials in water- and wastewater-treatment processes. Also termed "activated charcoal".
Activated Charcoal	→ Activated Carbon.
Activation Energy	The minimum potential energy that must be attained by a system for a reaction or process to take place at a significant rate. Catalysts

	usually function by providing a mechanism for a reaction that has a lower activation energy than does the uncatalyzed reaction.
Activator	Any agent that may be used in froth flotation to enhance, selectively, the effectiveness of collectors for certain mineral components. Example: sphalerite (ZnS) can be treated with copper sulfate (the activator), which adsorbs and bridges to ethyl xanthate (a collector), which in turn allows the sphalerite to be floated. → Froth Flotation.
Active Site	In adsorption, the specific regions of an adsorbent onto which a substance may adsorb. In catalysis, the site responsible for a particular reaction.
Active Surface Area	→ Fuchs Surface Area.
Active Surfactant	The primary surfactant in a detergent formulation. → Detergent.
Adagulation	The deposition of small, usually colloidal-sized particles onto larger size particles. Also termed Slime Coating.
Adamson, Arthur W. (1919–2003)	An American physical and surface chemist known for his work in inorganic photochemistry (he has been called "the father of inorganic photochemistry"), surface chemistry and chemical education. He is particularly known to colloid and interface scientists for his textbooks on physical chemistry and surface and colloid chemistry, especially "Physical Chemistry of Surfaces," which continued through six editions.
Adatom	An adsorbed atom.
Additive Electrolyte	→ Critical Coagulation Concentration.
Adhesion	(1) The attachment of one phase to another. → Work of Adhesion, → Adhesive. (2) The load causing failure of a joint, for example, of a glued joint between two materials. → Peel Test.
Adhesion Tension	An older term that referred to the wetting tension and/or the interfacial tension between solid and liquid phases. These usages have been discouraged to avoid confusion with the work of adhesion. *See also* Reference [4].
Adhesional Wetting	The process of wetting when a surface (usually solid), previously in contact with gas, becomes wetted by liquid. This term is sometimes used to describe wetting that includes the formation of an adhesional bond between the liquid and the phase it is wetting. → Wetting, → Spreading Wetting, → Immersional Wetting.
Adhesive	Any substance that enables or enhances mechanical adhesion, usually between solids. Example: glue.

Adjuvant	Chemicals that modify the effect of specific other chemicals while having few if any direct effects when they occur by themselves. Adjuvant surfactants increase the effects of biologically active compounds in agrochemical preparations. Their ability to perform this function is at least partly due to their ability to enhance wetting and spreading, which reduces the amount of active ingredient needed to be effective. Example: the use of nonionic surfactants in herbicide solutions sprayed on crops.
Admicelle	$\rightarrow$ Hemimicelle.
Admicellar Catalysis	Catalysis occurring in the admicellar (internal) region of admicelles adsorbed onto some medium. $\rightarrow$ Hemimicelle.
Admicellar Chromatography	The chromatographic separation of compounds as they pass through a medium containing media bearing admicelles. $\rightarrow$ Hemimicelle.
Adsolubilization	A surface analog of micellar solubilization in which adsorbed surfactant bilayers (admicelles) absorb solutes from solution. Example: the partitioning of sparingly soluble organic molecules from water into admicelles. *See* Reference [46].
Adsorbate	A substance that becomes adsorbed at the interface or into the interfacial layer of another material, or adsorbent. $\rightarrow$ Adsorption.
Adsorbent	The substrate material onto which a substance is adsorbed. $\rightarrow$ Adsorption.
Adsorbent Surface	$\rightarrow$ Adsorption Space.
Adsorptive Filtration	Filtration in which particles are attracted to and retained by filter elements due to electrostatic and/or dispersion forces.
Adsorption	The increase in quantity of a component at an interface or in an interfacial layer. In most usage it is positive, but it can be negative (depletion); in this sense negative adsorption is a different process from desorption. Adsorption may also denote the process of components accumulating at an interface.
Adsorption Capacity	The maximum amount of adsorbate that can be adsorbed by an adsorbent. The amount of adsorbed substance reached in a saturated solution, often where the solute is strongly adsorbed from a solution in which it has limited solubility.
Adsorption Complex	The combination of a (molecular) species that is adsorbed together with that portion of the adsorbent to which it is bound.
Adsorption Hysteresis	The phenomenon in which adsorption and desorption curves (isotherms) depart from each other.
Adsorption Isobar	$\rightarrow$ Adsorption Isotherm.

Adsorption Isostere	The function relating the equilibrium pressure to the equilibrium temperature for a constant value of the amount (or surface excess amount) of substance adsorbed by a specified amount of adsorbent.
Adsorption Isotherm	The mathematical or experimental relationship between the equilibrium quantity of a material adsorbed and the composition of the bulk phase, at constant temperature. The adsorption isobar is the analogous relationship for constant pressure, and the adsorption isostere is the analogous relationship for constant volume. $\rightarrow$ Langmuir Isotherm, $\rightarrow$ Freundlich Isotherm, $\rightarrow$ Polanyi Isotherm, $\rightarrow$ Gibbs Isotherm, $\rightarrow$ Brunauer-Emmett-Teller Isotherm, $\rightarrow$ Characteristic Isotherm, $\rightarrow$ Frenkel-Halsey-Hill Isotherm, $\rightarrow$ Temkin Isotherm.
Adsorption Site	$\rightarrow$ Active Site.
Adsorption Space	An interface is sometimes considered to comprise two regions, one containing a certain thickness of adsorbent; the other containing a certain thickness of the fluid phase. The former is termed the "surface layer of the adsorbent" (or adsorbent surface) and the latter is termed the "adsorption space".
Adsorption Surface Area	A surface area determined by calculation from experimental adsorption data, using an adsorption isotherm model. For example, the BET surface area is that calculated using the BET adsorption isotherm method.
Adsorptive Material	Material that is present in one or both of the bulk phases bounding an interface and capable of becoming adsorbed.
Advancing Contact Angle	The dynamic contact angle that is measured when one phase is advancing, or increasing, its area of contact along an interface while in contact with a third, immiscible phase. It is essential to state through which phase the contact angle is measured. $\rightarrow$ Contact Angle.
Advection	The transport of material solely by mass motion. In meteorology, an example is the transfer of heat by horizontal motion of the air. For flow in porous media, advective flow is without dispersion and results in the step appearance of chemical species at the downstream end of a control volume through which the species were flowing. In this case, the Darcy velocity alone is needed to predict the "advective" time of appearance of chemical species downstream.
AEAPS	Auger electron appearance potential spectroscopy. $\rightarrow$ Appearance Potential Spectroscopy.
Aeolotropic	$\rightarrow$ Anisotropic.

Aerated Emulsion	A foam in which the liquid consists of two phases in the form of an emulsion. Also termed "foam emulsion". Example: Whipped cream consists of air bubbles dispersed in cream, which is an emulsion. → Foam.
Aerating Agent	→ Foaming Agent.
Aeration	The dispersion and/or dissolution of air into a liquid.
Aerator	(1) Any machine used for preparing foams, especially in the food industry. In batch aerators, the gas is usually whipped into the liquid. In continuous aerators, a mixing head whips the gas into the liquid under pressure. In this case, the foam expands as it leaves the machine. → Oakes Mixer. (2) In environmental applications, any means for increasing the liquid-gas interface to promote either oxygen transfer into water (e.g., to enhance microbial reactions or oxidize compounds) or to enhance the mass transfer of volatile organic materials from the liquid phase to the gas phase.
Aerial Dispersion	→ Aerosol.
Aerocolloidal System	→ Aerosol.
Aerodynamic Atomizer	→ Air-Blast Atomizer.
Aerodynamic Diameter	The diameter of an imaginary spherical aerosol particle, having a density of $1.000\,\mathrm{g/cm^3}$, that has the same sedimentation velocity (in air) as the particle(s) under examination. → Equivalent Spherical Diameter.
Aerogel	A special kind of xerogel where the dried-out gel retains most of the original open structure. Example: silica gels that have been subjected to supercritical drying.
Aerosol	Colloidal dispersions of liquids or solids in a gas. Also referred to as aerocolloidal systems or aerial dispersions. → Aerosol of Liquid Droplets, → Aerosol of Solid Particles, → High-Dispersed Aerosols. Note that, in colloid science, aerosol does not refer to the consumer "aerosol spray cans", that is, the self-pressurized canisters containing hair sprays etc., although these products do deliver a very unstable aerosol when used. → Accumulation Aerosol, → Nanoaerosol, and → Nucleation Aerosol.
Aerosol Diffusion Charging	A method in which the Fuchs surface area of neutral aerosol particles is measured by passing them through an ion cloud and measuring the resulting aerosol charge.
Aerosol Knudsen Number	The ratio of the apparent mean free path of aerosol droplets or particles to the radius of those droplets or particles. For aerosol Knudsen numbers much greater than one, the aggregation rate for

dispersed aerosol species can be approximated by that for gas molecules. → High-Dispersed Aerosols, → Knudsen Number.

Aerosol of Liquid Droplets

A colloidal dispersion of liquid droplets in a gas. Distinctions may be made among aerosols of liquid droplets, such as fog, cloud, drizzle, mist, rain, and spray. → Atmospheric Aerosols of Liquid Droplets, *see* Table 5.

Aerosol of Solid Particles

A colloidal dispersion of solid particles in a gas. Distinctions may be made among aerosols of solid particles, such as fume and dust. → Atmospheric Aerosols of Solid Particles, *see* Table 6.

Aerosol Scavenging

The use of particles combined with diffusive, inertial, or gravitational processes to remove other, aerosol particles.

AES

→ Auger Electron Spectroscopy.

AFM

Atomic force microscopy. → Scanning Probe Microscopy, *see* Table 7.

Agar

A water-soluble mixture of polysaccharides derived from seaweeds. Agar is considered to be composed of three types of representative structures known as neutral agarose (or agaran), pyruvic acid acetal, and sulfated galactan. The combination of the latter two types is sometimes referred to as charged agar, or agaropectin. Agar sols can be quite viscous; can readily form gels; and may be used to stabilize certain suspensions, foams, and emulsions. Agar is used in many different applications including foods and medicines. *See also* Reference [47], → Seaweed Colloids, → Hydrocolloid.

Agaran

One of the kinds of polysaccharide structure that constitutes agar. Also termed "neutral agarose". → Agar.

Agglomerate

Usually, a cluster of particles, droplets, or bubbles that are weakly held together, such as by London-van der Waals forces for example. In some useage agglomerate is used to refer to a cluster of particles that is strongly bonded together, so it is preferable to specify whether strong or weak bonding is meant. → Aggregate.

Agglomerate Flotation

→ Oil-Assisted Flotation.

Agglomeration

The aggregation of particles, droplets, or bubbles in a dispersion. This term sometimes refers to a combination of aggregation and coalescence processes. → Agglomerate, → Spherical Agglomeration.

Agglutination

→ Aggregation.

Aggregate

A group of species, usually droplets, bubbles, particles, or molecules, that are held together in some way and not easily broken apart. A micelle can be considered an aggregate of surfactant molecules or ions. Context is important. In colloid and interface science this term refers to microscopic-sized species, whereas in soil science, aggregate is frequently used to refer to macroscopic crumbs in a soil. In the former case one may be concerned with colloidal stability, in the latter it is usually stability to mechanical disintegration. In nanotechnology aggregate refers to particles that are strongly bonded together, such as fused, sintered, or metallically-bonded particles. Since useage varies, it is preferable to specify whether strong or weak bonding is meant. $\rightarrow$ Agglomerate, $\rightarrow$ Marine Snow.

Aggregation

(1) The process of forming a group of droplets, bubbles, particles, or molecules that are held together in some way. This process is sometimes referred to interchangeably as "coagulation" or "flocculation", although in some usage these refer to aggregation at the primary and secondary minimum, respectively. The synonym agglutination has also been used (especially in biology). The reverse process is termed "deflocculation" or "peptization". $\rightarrow$ Primary Minimum, $\rightarrow$ Electrocoagulation, $\rightarrow$ Heterocoagulation, $\rightarrow$ Sweep Flocculation, $\rightarrow$ Polymer Charge Patch, $\rightarrow$ Polymer Bridging.

(2) For suspensions and emulsions, coagulation and flocculation are understood to represent two kinds of aggregation. In this case coagulation refers to the formation of compact aggregates, whereas flocculation refers to the formation of a loose network of particles. An example can be found in montmorillonite clay suspensions in which coagulation refers to dense aggregates produced by face-face oriented particle associations, and flocculation refers to loose aggregates produced by edge-face or edge-edge oriented particle associations. *See also* Reference [26].

Aggregation Half-Life

$\rightarrow$ Aggregation Time.

Aggregation Number

The number of surfactant molecules or ions composing a micelle. Example: The aggregation number for dodecyl sulfate ions in water is about 70.

Aggregation Time

For perikinetic aggregation, the Smoluchowski equation predicts that the total dispersed species concentration decreases to one-half of its original concentration after a characteristic time, termed the aggregation time (or coagulation time). This time is approximately the average interval between collisions for a given species. Also termed the aggregation half-life (or coagulation half-life). $\rightarrow$ Limiting Collision Efficiency.

Aggregative Stability

Stability against aggregation. Used in Russian colloid literature.

Aging	The properties of many colloidal systems may change with time in storage. (1) Petroleum: Aging in crude oils can refer to changes in composition due to oxidation, precipitation of components, bacterial action, or evaporation of low-boiling components. (2) Emulsions and foams: Aging in emulsions or foams can refer to any changes of aggregation, coalescence, creaming or chemical. Aged emulsions and foams frequently have larger droplet or bubble sizes. (3) Suspensions: Aging in suspensions usually refers to aggregation, that is, coagulation or flocculation. It is also used to describe the process of recrystallization, in which larger crystals grow at the expense of smaller ones, that is, Ostwald ripening.
Agitator	A general term referring to mechanical mechanisms that mix and recirculate colloidal dispersions within vessels. The mechanisms may include propellors, paddles, turbines, or shaking devices.
Agitator Ball Mill	A machine for the comminution, or size reduction, of minerals or other materials. Such machines crush the input material by wet grinding in a cylindrical rotating bin containing grinding balls. These mills can produce colloidal-size particles.
Air-Blast Atomizer	Device for making aerosols of liquid droplets by ejecting compressed gas at high velocity into a liquid stream emerging from a nozzle. Example: paint spray guns. Also termed aerodynamic atomizer or venturi atomizer.
Air Drilling Fluid	Air, when used as an oil- and gas-well drilling fluid. An air drilling fluid may contain a small amount of water, in which case a more specific term is mist drilling fluid. If the water also contains a foaming agent (surfactant), then the more specific term is foam drilling fluid. Gases other than air are sometimes used, such as nitrogen or natural gas. → Foam Drilling Fluid, → Stable Foam, → Stiff Foam.
Airless Spraying	A method for atomizing and spraying a liquid, suspension, or emulsion by high pressure, without using compressed gas at the spray nozzle. Also termed "hydraulic spraying". Airless spraying is used for paints and urethanes among others.
Aitken Counter	An instrument used for counting the particles in aerosols of solid particles. Before counting, moist air is quickly expanded to create supersaturation, which in turn leads to condensation of water vapour onto the particle surfaces. This produces particle-containing droplets that are more easily viewed, and counted, than the original aerosol particles. Also termed Aitken dust counter, or Aitken nucleus counter. *See also* Table 6, → Aerosol of Solid Particles, → Atmospheric Aerosols of Solid Particles.

Aitken Nuclei	Aerosol particles having diameters smaller than 0.2 micrometres. Example: some combustion by-products. *See also* Table 6, $\rightarrow$ Aerosol of Solid Particles.
Alcohol Resisting Aqueous Film Forming Foam	(AFFF-AR) A fire-extinguishing foam formulated specifically for alcohol, polar solvent, and hydrocarbon fires. $\rightarrow$ Fluoroprotein Foam, $\rightarrow$ Film Forming Fluoroprotein Foam, $\rightarrow$ Aqueous Film Forming Foam, $\rightarrow$ Fire Extinguishing Foam.
Algin	Any of the salt forms of alginic acid, which is a polysaccharide derived from seaweeds. Most of the salt forms are highly water-soluble. Also termed alginate. Algin sols can be quite viscous, can readily form gels, and can be used to stabilize certain suspensions and emulsions. Algin is used in many different applications including food processing. *See also* Reference [35], $\rightarrow$ Seaweed Colloids.
Alginate	$\rightarrow$ Algin.
Alginic Acid	$\rightarrow$ Algin.
Alkane Carbon Number	(ACN) $\rightarrow$ Equivalent Alkane Carbon Number.
Allotropy	Referring to different crystal structures of the same element. Polymorphism refers to different crystal structures of the same compound.
Amicron	An older particle-size range distinction no longer in use. $\rightarrow$ Micrometre, $\rightarrow$ Micron.
Amonton's Law	A description of friction that states that the coefficient of friction is given by the frictional force divided by the load normal to the direction of motion along the surfaces. $\rightarrow$ Friction, $\rightarrow$ Lubrication.
Amott-Harvey Test	$\rightarrow$ Amott Test.
Amott Test	A measure of wettability based on a comparison of the amounts of water or oil imbibed into a porous medium spontaneously and by forced displacement. Amott test results are expressed as a displacement-by-oil (δ_o) ratio and a displacement-by-water ratio (δ_w). In the Amott-Harvey test, a core is first prepared at irreducible water saturation. The Amott-Harvey relative displacement (wettability) index is then calculated as $\delta_w - \delta_o$, with values ranging from -1.0 for complete oil-wetting to 1.0 for complete water-wetting. *See also* Reference [48], $\rightarrow$ Wettability, $\rightarrow$ Wettability Index.
Amott Wettability Index	$\rightarrow$ Amott Test.
Amphipathic	Having both lyophilic and lyophobic groups (properties) in the same molecule, as in the case of surfactants. Also referred to as being "amphiphilic".
Amphiphilic	$\rightarrow$ Amphipathic.

Ampholytic Surfactant	$\rightarrow$ Amphoteric Surfactant.
Amphoteric Surfactant	A surfactant molecule for which the ionic character of the polar group depends on solution pH. Also termed ampholytic surfactant. For example, Lauramidopropyl betaine $C_{11}H_{23}CONH(CH_2)_3N^+(CH_3)_2CH_2COO^-$ is positively charged at low pH but is electrically neutral, having both positive and negative charges at intermediate pH. Other combinations are possible, and some amphoteric surfactants are negatively charged at high pH. $\rightarrow$ Zwitterionic Surfactant.
Analytical Ferrography	A method for optical analysis of particulates produced during the wearing down of machinery. It involves magnetic precipitation of wear debris from a fluid sample onto a chemically treated microscope slide. A magnetic field gradient causes the particles to deposit in a distribution with respect to size and mass over the ferrogram, which is optically examined. *See* Reference [49].
Ancillaries	The non-surface active, complementary components in a detergent formulation. $\rightarrow$ Detergent.
Andreason Pipet	A graduated cylinder having provision for withdrawing subsamples from the bottom [8]. Used to study sedimentation in the determination of particle sizes. $\rightarrow$ Stokes' Law.
Angle of Incidence	The angle between an incident beam and the normal to a surface. Also termed the tilt of a target.
Angle-Resolved Photoemission Extended Fine Structure	(ARPEFS) $\rightarrow$ Angle-Resolved Photoemission Spectroscopy.
Angle-Resolved Photoemission Spectroscopy	(ARPES) Includes the family of angle-resolved photoemission techniques used for surface structure determination, including angle-resolved photoemission extended fine structure (ARPEFS), ultraviolet (ARUPS), X-ray (ARXPS), and X-ray diffraction (ARXPD). *See* Reference [50].
Angle-Resolved Ultraviolet Photoemission Spectroscopy	(ARUPS) $\rightarrow$ Angle-Resolved Photoemission Spectroscopy.
Angle-Resolved X-Ray Photolectron Diffraction	(ARXPD) $\rightarrow$ Angle-Resolved Photoemission Spectroscopy.
Angle-Resolved X-Ray Photoemission Spectroscopy	(ARXPS) $\rightarrow$ Angle-Resolved Photoemission Spectroscopy.
Animal Glue	$\rightarrow$ Protein Colloid.
Anion-Exchange Capacity	The capacity for a substrate to adsorb anionic species while simultaneously desorbing (exchanging) an equivalent charge quantity of other anionic species. Example: This property is

sometimes used to characterize clay minerals that often have very large cation-exchange capacities but may also have significant anion-exchange capacities. → Ion Exchange.

Anionic Surfactant

A surfactant molecule that can dissociate to yield a surfactant ion whose polar group is negatively charged. Example: sodium dodecyl sulfate, $CH_3(CH_2)_{11}SO_4^-Na^+$.

Anisokinetic Sampling

→ Isokinetic Sampling.

Anisotropic

A material that exhibits a physical property, such as light transmission, differently in various directions. Sometimes termed "aeolotropic".

Anneal

The process of heating a solid material to a temperature close to, but lower than, its melting point to reduce internal stresses and strengthen the material.

Anode

An electrode at which a net positive current flows. The predominant chemical reaction here is oxidation.

Anodic Oxidation

(Nanotechnology) → Scanning Probe Surface Patterning.

Anomalous Viscosity

An older term for "non-Newtonian viscosity" [51]. → Non-Newtonian Flow, → Non-Newtonian Fluid.

Anomalous Water

→ Polywater.

Antagonistic Electrolyte

→ Critical Coagulation Concentration.

Anti-Bubbles

A dispersion of liquid-in-gas-in-liquid wherein a droplet of liquid is surrounded by a thin layer of gas that in turn is surrounded by bulk liquid. Example: In an air-aqueous surfactant solution system this dispersion would be designated as water-in-air-in-water, or W/A/W, in fluid film terminology. A liquid-liquid analogy can be drawn with the structures of multiple emulsions. *See also* Reference [52], → Fluid Film.

Antielectrostatic Agent

A surfactant formulation that can be applied to a fabric or fibres to reduce the buildup of static electricity. Examples: alkyl sulfonates and alkyl phosphates.

Antifoaming Agent

Any substance that acts to reduce the stability of a foam; it can also act to prevent foam formation. Terms such as "antifoamer" or "foam inhibitor" specify the prevention of foaming, and terms such as "defoamer" or "foam breaker" specify the reduction or elimination of foam stability. Example: Poly(dimethylsiloxane)s, $(CH_3)_3SiO$ $[(CH_3)_2SiO]_xR$, where R represents any of a number of organic functional groups. Antifoamers can act by any of a number of mechanisms.

Anti-redeposition Agent	A component in a detergent formulation that acts to help prevent redeposition of dispersed dirt or grease. Example: carboxymethyl cellulose. *See also* Reference [53], $\rightarrow$ Detergent.
Antistatic Agent	$\rightarrow$ Antielectrostatic Agent.
Antithixotropy	$\rightarrow$ Rheopexy.
Antonow's Rule	An empirical rule for the estimation of interfacial tension between two liquids as the difference between the surface tensions of each liquid. An analogous form gives the solid/liquid interfacial tension as the difference between the liquid and solid surface tensions. Even for pure liquids this rule is seldom very accurate. *See* Reference [54] and Table 8.
A/O/W	An abbreviation for a fluid film of oil between air and water. Usually designated W/O/A. $\rightarrow$ Fluid Film.
APD	Azimuthal photoelectron diffraction. $\rightarrow$ Photoelectron Diffraction.
Aphrons	$\rightarrow$ Colloidal Liquid Aphrons, $\rightarrow$ Microgas Emulsions.
API Gravity	A measure of the relative density (specific gravity) of petroleum liquids. The API gravity, in degrees, is given by $^\circ API = (141.5/$relative density$) - 131.5$, where the relative density at temperature t ($^\circ$C) = (density at t)/(density of water at 15.6 $^\circ$C).
Apolar	Description applied to materials or surfaces that have no polar nature.
Apparent Viscosity	Viscosity determined for a non-Newtonian fluid without reference to a particular shear rate for which it applies. Alternatively, viscosity determined for a non-Newtonian fluid, but at only one (usually high) shear rate. Such viscosities are usually determined by a method strictly applicable to Newtonian fluids only. *See* Table 4. $\rightarrow$ Effective Viscosity.
Appearance Potential Spectroscopy	(APS) A technique related to photoelectron spectroscopy and also used for the determination of surface composition. The surface is scanned with an electron beam of varying energy, which causes the ejection of inner electrons from the surface atoms. The intensities of the beams of ejected electrons are determined (X-ray or Auger electrons). The terms Auger electron appearance potential spectroscopy (AEAPS) and soft X-ray appearance potential spectroscopy (SXAPS, also termed Appearance Potential X-Ray Photoemission Spectroscopy, APXPS) are used to distinguish modes in which Auger electrons or photons, respectively, are emitted. *See also* Table 7.

Appearance Potential X-Ray Photoemission Spectroscopy	(APXPS) → Appearance Potential Spectroscopy.
APS	→ Appearance Potential Spectroscopy.
APXPS	→ Appearance Potential Spectroscopy.
Aquacolloid	→ Hydrocolloid.
Aquadag	A commercial colloidal graphite product (a suspension of colloidal graphite in water). → Acheson, Edward.
Aquasol	→ Hydrosol.
Aqueous Emulsion	An emulsion having an aqueous continuous phase.
Aqueous Film Forming Foam	(AFFF) A fire extinguishing foam based on blended hydrocarbon and fluourocarbon surfactants; a rapidly spreading foam used, for example, on oil platform helidecks. → Fluoroprotein Foam, → Film Forming Fluoroprotein Foam, → Alcohol Resisting Aqueous Film Forming Foam, → Fire Extinguishing Foam.
Areal Elasticity	→ Film Elasticity.
ARPEFS	Angle-Resolved Photoemission Extended Fine Structure. → Angle-Resolved Photoemission Spectroscopy.
ARPES	→ Angle-Resolved Photoemission Spectroscopy.
Artificial Atoms	→ Quantum Dot.
ARUPS	Angle-Resolved Ultraviolet Photoemission Spectroscopy. → Angle-Resolved Photoemission Spectroscopy.
ARXPD	Angle-Resolved X-ray Photolectron Diffraction. → Angle-Resolved Photoemission Spectroscopy.
ARXPS	Angle-Resolved X-ray Photoemission Spectroscopy. → Angle-Resolved Photoemission Spectroscopy.
Arrhenius, Svante (August) (1859–1927)	A Swedish chemist and physicist who (with van't Hoff and Ostwald) helped establish the discipline of physical chemistry. He did considerable work on the electrical conductivity of solutions and the dissociation of salts, and he made contributions to electrochemistry. He was awarded the Nobel prize in chemistry (1903) for his electrolytic theory of dissociation.
Aspect Ratio	The ratio of the largest Feret's diameter of a species to the smallest perpendicular diameter. → Elongation Shape Factor.
Asperities	Microscopic projections on metal surfaces.
Asphalt	A naturally occurring hydrocarbon that is a solid at reservoir temperatures. An asphalt residue can also be prepared from

heavy (asphaltic) crude oils or bitumen, from which lower boiling fractions have been removed, and/or which are highly oxidized.

Asphaltene

A high-molecular-mass, polyaromatic component of some crude oils that also has high sulfur, nitrogen, oxygen, and metal contents. In practical work asphaltenes are usually defined operationally by using a standardized separation scheme. One such scheme defines asphaltenes as those components of a crude oil or bitumen that are soluble in toluene but insoluble in n-pentane.

Assembler

(Nanotechnology) A device for conducting a mechanical action on the nanoscale. Nanotechnology assemblers modelled on devices from the macroscopic world include pumps, bearings, drive shafts, bearings, and so on. A simple example is a device that can pick-up an atom and move it to another position, such as with an atomic force microscope. Such a device has also been termed a nanoassembler, or a nanocrane. The precise positioning of reactive molecules in order to control chemical reactions is termed positional synthesis. → Breadbox Assembler.

Association Colloid

A dispersion of colloidal-sized aggregates of small molecules; it is lyophilic. Also termed "self-assembling colloid". Example: micelles of surfactant molecules or ions in water.

Associative Polymers

→ Hydrophobically Associating Polymers.

Asymmetric Film

→ Film.

Atactic Polymer

In polymers with monomer units of the form (CH_2CHR), the hydrogen atoms and R groups may tend to align differently with respect to an imaginary plane containing the carbon atom chain. In an atactic polymer the orientation is random; in a syndiotactic (or syntactic) polymer the R groups alternate from side to side; in an isotactic polymer all the R groups lie on one side and all the hydrogen atoms lie on the other. → Tacticity.

Atmospheric Aerosols of Liquid Droplets

Atmospheric colloidal dispersions of liquids in gas. Distinctions are made among fog, cloud, drizzle, and rain, depending upon droplet sizes and based on whether the droplets would be large enough to fall to the earth's surface before completely evaporating. Droplets small enough to evaporate before reaching the ground fall into the fog and cloud ranges, less than about 100 µm. Droplets greater than about 100 µm are large enough to reach the ground before evaporating and fall into the rain category. Droplets with diameters of about 100 µm are termed drizzle and correspond to a stage in the development of rain through the coalescence of cloud droplets. *See* Table 5.

Atmospheric Aerosols of Liquid Particles	Atmospheric colloidal dispersions of solids in gas. Distinctions are made among, for example, fume and smoke (0.01–1 µm), dust (0.5–100 µm), and ash (1–500 µm). → Aitken Nuclei, *see* Table 6.
Atomic Force Microscopy	(AFM) → Scanning Probe Microscopy, *see* Table 7.
Atomization	A method for the preparation of fine solid particles, by spraying a molten material, solution, or suspension in a way that causes the dispersed droplets to break-down and solidify into a powder or an aerosol of solid particles. The particles prepared in this way are typically of the order of a few micrometres in diameter. Also termed nebulization. → Electrohydrodynamic Atomization.
Atterberg Limits	A group of (originally) seven limits of soil consistency, or relative ease with which material can be deformed or made to flow. The only Atterberg limits that are still in common use are the liquid limit, plastic limit, and plasticity number. *See* References [55, 56].
Attractive Potential Energy	→ Gibbs Energy of Attraction.
Attrition	The reduction of particle size by erosion due to friction and wear. Also termed ultrafine grinding or nanosizing. → Ablation, → Comminution.
Auger Electron Appearance Potential Spectroscopy	(AEAPS) → Appearance Potential Spectroscopy.
Auger Electron Spectroscopy	(AES) A technique used for the determination of surface composition by scanning the surface with an electron beam. The beam ionizes surface atoms by ejecting inner-shell electrons. Electron transfer from outer electron shells will result in the emission of energy, either as characteristic X-rays, or in the ejection of a second outer-shell electron (Auger electron). Auger electrons have energies characteristic of the atoms from which they were ejected. *See also* Table 7. *See* Reference [57] for specific terms in AES.
Autophobicity	→ Autophobing.
Autophobing	A phenomenon involving the spreading of a surfactant solution over a surface in which surfactant adsorption on the surface causes a wettability reversal, which in turn causes the advancing liquid front to halt and then retreat. Example: a drop of cationic surfactant solution that initially spreads over a negatively charged hydrophilic surface, but for which surfactant adsorption on the surface causes the wettability of the surface to reverse to hydrophobic, which in turn causes the droplet to dewet and recede. In the case of liquid films rather than droplets, autophobing can cause the formation of dewetted areas, or holes, in the liquid film. → Spreading Coefficient.

Average Agglomeration Number	(AAN) An estimate of the degree of agglomeration of particles in a suspension, such as the average number of particles in an aggregate.
A/W/A	An abbreviation for a fluid film of water in air. $\rightarrow$ Fluid Film.
A/W/O	An abbreviation for a fluid film of water between air and oil phases. Also termed pseudoemulsion film. Usually designated O/W/A. $\rightarrow$ Fluid Film.
Azimuthal Photoelectron Diffraction	(APD) $\rightarrow$ Photoelectron Diffraction.

B

Backscattered Electrons	In electron spectroscopy, such as AES, electrons originally from the incident beam that are emitted after interaction with a target.
Backscattering	$\rightarrow$ Light Scattering.
Bacteria	Small single-celled microorganisms, typically between 200 nm and 2000 nm in diameter or length. They can form stable colloidal dispersions in water because they tend to have a net negative surface charge.
Bamboo-Concentric Multi-Wall Nanotube	(bc-MWNT) $\rightarrow$ Carbon Nanotube.
Bamboo Foam	A descriptive term for foam that is confined in a narrow diameter tube, or channel, so that the foam films (lamellae) are arranged in more or less parallel alignment.
Bamboo-Herringbone Multi-Wall Nanotube	(bh-MWNT) $\rightarrow$ Carbon Nanotube.
Bancroft, Wilder Dwight (1867–1953)	An American physical chemist who made many contributions to colloid chemistry, including two books and the founding of the Journal of Physical Chemistry. His empirical generalization, of which phase will be the continuous phase in an emulsion, is known as Bancroft's Rule. *See* References [58, 59, 130].
Bancroft's Rule	An empirical generalization that predicts that the continuous phase in an emulsion will be the phase in which the emulsifying agent is most soluble. An extension for solid particles acting as emulsifying agents predicts that the continuous phase will be the phase that preferentially wets the solid particles. $\rightarrow$ Hydrophile-Lipophile Balance.

Dictionary of Nanotechnology, Colloid and Interface Science. Laurier L. Schramm
Copyright © 2008 WILEY-VCH Verlag GmbH & Co. KGaA, Weinheim
ISBN: 978-3-527-32203-9

Barrett-Joyner-Halenda Theory	(BJH Theory) In the physical adsorption of gas molecules beyond monolayer formation and filling of pores via capillary condensation once critical relative pressures are achieved, BJH theory allows one to calculate pore sizes from the equilibrium gas pressures. From the BJH isotherms one can obtain pore size distributions. *See* References [60, 61].
Bartsch Test	A static foam stability test involving the carefully prescribed shaking of a known volume of solution in a sealed container, such as a bottle, for a specified period of time and/or a specified number of shakes, to simulate conditions of relatively low shear. → Ross-Miles Test, → Static Foam Test.
Basic Sediment and Water	That portion of solids and aqueous solution in an emulsion that separates out on standing or is separated by centrifuging in a standardized test method. Basic sediment may contain emulsified oil as well. Also referred to as BS&W, "BSW", Bottom Settlings and Water, and Bottom Solids and Water.
Batch Mixer	A type of processing equipment in which the entire amount of material to be used is put into the mixer and mixed for a definite period, with multiple recirculation of material through the mixing zone, in contrast to what happens in a continuous mixer. After the mixing period the whole amount of material is removed from the mixer.
Batch Treating	In oil production or processing, the process in which emulsion is collected in a tank and then broken in a batch. This process is used as opposed to continuous or flow-line treating of emulsions.
bc-MWNT	Bamboo-Concentric Multi-Wall Nanotube. → Carbon Nanotube.
Beaker Test	→ Bottle Test.
Bed Knives	The stationary cutting blades in a cutting mill machine for comminution.
Beer-Lambert Law	→ Absorbance.
Bénard Cells	Originally, convection cells that appear spontaneously in a liquid layer when heat is applied from below. More generally, the cellular flow patterns that appear due to thermal gradients during the uneven drying of a film coating. This is an example of Marangoni flow as the thermal gradients create surface tension gradients.
Beneficiation	In mineral processing, any process that results in a product having an improved desired mineral content. Example: froth flotation.
Berzelius, Jöns Jacob, Baron (1779–1848)	A Swedish chemist known for developing the first major systematization of 19th century chemistry, the first accurate table of relative atomic masses, and the system of chemical notation that is still in

use today. He also discovered or co-discovered the elements selenium, cerium, silicon, and thorium. His publications in the early 1800s described the preparation and analysis of thousands of compounds. Berzelius coined the terms catalytic, halogens, organic chemistry, and isomerism, and his 1803 "Textbook of Chemistry" (1803) was the standard of its time.

BET Analysis Analysis of gas adsorption data, using the BET model, in order to determine the surface area of a solid material. $\rightarrow$ Brunauer-Emmett-Teller Isotherm.

BET Isotherm $\rightarrow$ Brunauer-Emmett-Teller Isotherm.

BET Surface Area $\rightarrow$ Adsorption Surface Area.

bh-MWNT Bamboo-Herringbone Multi-Wall Nanotube. $\rightarrow$ Carbon Nanotube.

Bicontinuous System A two-phase system in which both phases are continuous phases. For example, a possible structure for middle-phase microemulsions is one in which both oil and water phases are continuous throughout the microemulsion phase. An analogy can be drawn from the structure of porous and permeable rock in which both the mineral phase and the pore or throat channels can be continuous at the same time. A bicontinuous microemulsion is sometimes termed a sponge phase. $\rightarrow$ Middle-Phase Microemulsion.

Bilayer $\rightarrow$ Bimolecular Film.

Biliquid Foam A concentrated emulsion of one liquid dispersed in another liquid.

Bimolecular Film A membrane that separates two aqueous phases and is composed of two layers of polar organic molecules, such as surfactants or lipids. These molecules are oriented with their hydrocarbon groups in the two molecular layers towards each other and the polar groups facing the respective aqueous phases. $\rightarrow$ Vesicle, $\rightarrow$ Black Lipid Membrane.

Binder The film-forming component in a surface coating or paint. Binders provide cohesion among pigment particles and adhesion to the intended surface. Binders have consisted of a wide range of materials from natural gums and resins to synthetic polymers. Synonyms include resin and vehicle. Also termed resin or vehicle.

Bingham Plastic Fluid $\rightarrow$ Plastic Fluid.

Biochip $\rightarrow$ Biomedical Microelectromechanical Systems.

Biocolloidal Dispersion A colloidal dispersion in which the dispersed phase is of biological origin. Example: a dispersion of lipid particles.

Bioflip $\rightarrow$ Biomedical Microelectromechanical Systems.

Biomedical Microelectro-mechanical Systems	(BioMEMS) A subset of microelectromechanical systems (MEMS) devices for applications in biomedical research and biomedicine. MEMS devices involve the integration of mechanical structures with microelectronics and are designed for specific purposes such as sensors and process controls. BioMEMS devices generally have at least one dimension in the 100 to 200 nm range. BioMEMS functions include microsensing, microactuating, microassaying, micromoving, and microdelivery. The science of MEMS includes the materials science aspects. BioMEMS examples include micro-fluidic chips (also known as microfluidic chips, bioflips, biochips) used in biosensors. *See* Reference [62].
Biomedical Nanoelectro-mechanical Systems	(BioNEMS) The nanoscale equivalent of Biomedical Microelec-tromechanical Systems (BioMEMS). An example of a BioNEMS device is a nanobiosensor comprising a microarray of silicon nanowires, or carbon nanotubes, to selectively bind and detect one or a few biological molecules, using micro- or nanoelectronics to detect the slight electrical charge caused by such binding.
BioMEMS	→ Biomedical Microelectromechanical Systems.
Bionanotechnology	→ Nanobiotechnology.
BioNEMS	Biomedical Nanoelectromechanical Systems. → Biomedical Microelectromechanical Systems.
Bioremediation	The use of biological agents in the treatment and reclamation of contaminated soils or waters.
Bipolar	A substance having electron-donor as well as electron-acceptor properties. This feature has an influence on surface tension. Bipolar is not the same as dipolar. → Dipole.
Birefringent	A material that has different refractive indices in different direc-tions. Example: liquid crystals. Birefringence can be induced by applied stress in glass and gels.
Bitter Colloid Method	A means of observing the influence of external magnetic fields on the structure of a ferrofluid. In the ferrofluid the equilibrium between steric and electrostatic repulsive forces and van der Waals attractive forces and magnetic forces is influenced by any external magnetic field, which leads to differences in colloid particle densities, and agglomeration, in the ferrofluid. Such concentration changes and agglomerates can be observed if the continuous liquid phase is removed, leaving "Bitter patterns." *See* Reference [63]. → Ferrofluid.
Bitumen	A naturally occurring viscous hydrocarbon with a viscosity greater than 10,000 mPa·s at ambient deposit temperature, and a density greater than 1000 kg/m^3 at 15.6 °C [64–66]. Bitumen is also sometimes used to refer to viscous residue from the vacuum

distillation of crude oil. In either usage, in addition to high-molecular-mass hydrocarbons, bitumen contains appreciable quantities of sulfur, nitrogen, oxygen, and heavy metals. It is also a colloidal dispersion of asphaltenes.

BJH Theory
$\rightarrow$ Barrett-Joyner-Halenda Theory.

Black Film
Fluid films yield interference colours in reflected white light that are characteristic of their thickness. At a thickness of about 0.1 μm the films appear white and are referred to as "silver films". At reduced thicknesses they first become grey and then black (black films). Among thin equilibrium (black) films, one may distinguish those that correspond to a primary minimum in interaction energy, typically at thicknesses of about 5 nm (Newton black films) from those that correspond to a secondary minimum, typically at thicknesses of about 30 nm (common black films).

Blacking
Powdered graphite.

Black Lead
$\rightarrow$ Graphite.

Black Lipid Membrane
A bimolecular film in which the molecules composing the membrane film are lipid molecules. The term "black" refers to the fact that these films appear black when illuminated (no apparent interference colors). $\rightarrow$ Bimolecular Film, $\rightarrow$ Black Film.

Blender Test
An empirical test in which an amount of potential foaming agent is added into a blender containing a specified volume of liquid to be foamed. After blending at a specified speed and for some specified time, the blending is halted and the extent (volume) of foam produced is measured immediately and after a period of quiescent standing. This test has many variations. $\rightarrow$ Bottle Test.

Blinder
A chemical, usually a polymer, that is used to adsorb onto undesired mineral components that would otherwise adsorb collectors. Example: the addition of carboxymethyl cellulose (CMC) to a potash slurry where the CMC adsorbs onto insoluble minerals, preventing amine collector from being adsorbed, and preventing flotation of the undesired minerals. Also termed blinding agent; the process is termed blinding. $\rightarrow$ Froth Flotation.

BLM
$\rightarrow$ Black Lipid Membrane.

Block Copolymer
A polymer composed of two kinds of monomers in which the repeating unit consists of a chain, or block, of several units of each monomer type.

Blodgett, Katharine (Burr) (1898–1979)
An American industrial physicist and physical chemist who is known for her work in surface chemistry. She is especially known for her work in monomolecular and multilayer films (termed

"Blodgett films") and her invention of non-reflecting ("invisible") glass, which is used in optical instruments. *See* References [67, 68].

Blossom Thinning

A chemical, as opposed to manual, thinning procedure in which a surfactant or other chemical agent is applied to fruit trees while in bloom. This stops fertilization (probably usually by inhibiting photosynthesis) and reduces the number of fruits produced, which in turn increases the size of the fruits.

Blowing Agent

A chemical agent in a formulation that provides gas during processing. The gas may result from heating or from a chemical reaction. Example: Water reacts with isocyanate material to produce carbon dioxide gas in one process for making polyurethane (solid) foam.

Blue Emulsion

An emulsion in which the dispersed droplet diameters are less than about 200 nm so that they scatter blue light, giving the emulsion a blue glimmer. This property is used in some cosmetic emulsions. → Rayleigh Scattering.

Boger Fluid

A viscoelastic liquid whose viscosity does not change under applied shear. That is, although viscoelastic, such liquids are not, or at least substantially not, shear-thinning or shear-thickening. Boger fluids typically comprise a dilute solution of high molar mass polymer in a viscous liquid. *See* Reference [69]. → Viscoelastic, → Non-Newtonian Fluid.

Bola-Amphiphile

A surfactant that contains two hydrophilic (polar) head-groups, separated by a hydrophobic spacer, in the same molecule. Also termed bolytes. Depending on the length and flexibility of the spacer chain, bola-amphiphiles can self-assemble into aggregates having different morphologies, including spheres, cylinders, lamellae, discs, and vesicles.

Boltzmann Constant

A fundamental physical constant occurring in statistical formulations of classical and quantum physics.

Boltzmann, Ludwig (Eduard) (1844–1906)

An Austrian physicist who did important work on the kinetic theory of gases and established the principle of the equipartition of energy (Boltzmann's Law). He laid the foundations of statistical mechanics, which explains how the properties of atoms determine the physical properties of matter. The *Boltzmann constant* is a fundamental physical constant.

Boltzmann Equation

A distribution function giving the number of noninteracting or weakly interacting species with given energy, at a particular tremperature. In colloid science the Boltzmann equation gives the local concentrations of ions, in an electric double layer, in terms of the local electric potential. → Poisson–Boltzmann Equation.

Bond Number	A dimensionless ratio of gravitational or buoyancy forces (usually) to surface tension forces. The Bond number is given by $Bo = (\rho g L^2)/\gamma$ where ρ is the density, g is the acceleration due to gravity, L is the characteristic radius, and γ is the surface tension. Gravity effects tend to dominate in a system having a high Bond number whereas surface tension effects tend to dominate in a system having a low number Bond number ($Bo < 1$). A synonym for Bond number is Eötvös number.
Born Equation	An equation giving the Gibbs free energy of solvation of an ion.
Born, Max (1882–1970)	A German theoretical physicist and co-winner of the Nobel prize in physics (1954) for his work in quantum mechanics and his statistical interpretation of the wavefunction. He is remembered in colloid science through "Born repulsion", an important factor in colloid stability. Other eponyms include the "Born equation" (ion solvation) and a number of interatomic pair-potential equations, "Born-Mayer-Bohr potential," "Born-Mayer-Huggins potential," and "Born-Mayer potential." *See* Reference [70].
Böttcher Equation	For predicting the relative permittivity of dispersions. *See* Table 9.
Bottle Test	(1) Emulsions: An empirical test in which varying amounts of a potential demulsifier or coagulant are added into a series of tubes or bottles containing subsamples of an emulsion or other dispersion that is to be broken or coagulated. After some specified time the extent of phase separation and appearance of the interface separating the phases are noted. This test has many variations. For emulsions, in addition to the demulsifier, a diluent can be added to reduce viscosity. In the centrifuge test, centrifugal force can be added to speed up the phase separation. Other synonyms include jar test, beaker test.
	(2) Foams: An empirical test in which an amount of potential foaming agent (or even defoaming agent) is added into a bottle containing a specified volume of liquid to be foamed. The bottle is shaken in a specific manner, for some specified time, after which the shaking is halted and the extent (volume) of foam produced is measured immediately and after a period of time of quiescent standing. This test has many variations. → Blender Test.
	(3) Water treatment: A standard test method in which either the coagulant dosage is varied or the solution pH is varied for a given coagulant dosage, to optimize the coagulation of solids. Frequently termed "jar test".
Bottom Settlings and Water	→ Basic Sediment and Water.
Bottom Solids and Water	→ Basic Sediment and Water.

Bottom-Up Processing	An additive, building-up process by which nanostructures are made from constituent atoms and/or molecules. $\rightarrow$ Top-Down Processing.
Boundary Lubrication	$\rightarrow$ Lubrication.
Boussinesq Number	A measure of the ratio of interfacial and bulk viscous effects in a thinning foam film: $B_o = (\eta^s + \kappa^s)/(\eta R_f)$ where η^s is the surface shear viscosity, κ^s is the surface dilational viscosity, η is the bulk liquid viscosity, and R_f is the thin film radius.
Bragg, Sir William (Henry) (1862–1942)	A British physicist known for his work in solid-state physics. He was a joint winner (with his son Sir Lawrence Bragg) of the Nobel prize (1915) in physics for his research on the determination of crystal structures by means of X-ray diffraction. Bragg's Law is fundamental to X-ray crystallography. He was also a president of the Royal Society 1935–1940.
Bragg, Sir (William) Lawrence (1890–1971)	An Australian-born British physicist and X-ray crystallographer. He was joint winner (with his father, Sir William Bragg) of the Nobel prize (1915) in physics for his research on the determination of crystal structure by means of X-ray diffraction. Bragg's Law is fundamental to X-ray crystallography.
Bragg's Law	The relation between the spacing of atomic planes in crystals and the angles of incidence at which these planes produce the most intense reflections of electromagnetic radiations, such as X-rays or gamma rays. Used in X-ray crystallography.
Bragg's Rule	An empirical rule that relates the stopping cross-section of a compound to the sum of the products of the elemental stopping cross sections of each constituent and its atomic fraction. Used in energetic ion analysis. *See* Reference [57].
Branched-Chain Aggregate	$\rightarrow$ Dendritic.
Breadbox Assembler	(Nanotechnology) A conceptual device comprising many assemblers, replicators and nanocomputers, and capable of constructing larger than nanoscale objects and devices.
Breaking	The process in which an emulsion or foam separates. Usually coalescence causes the separation of a macrophase, and eventually the formerly dispersed phase becomes a continuous phase, separate from the original continuous phase.
Bredig Arc Method	A method in which metal particles are dispersed by passing an electric current between two wires (forming the arc) immersed in a liquid.
Bremsstrahlung	Electromagnetic radiation produced by the deceleration of incident electrons, when deflected by another charged particle, such as an atomic nucleus in a target material.

Brewster Angle Microscope	A light microscope based on the reflectivity of light close to the Brewster angle (the angle for which reflectivity of plane polarized light from an interface vanishes or is at a minimum). This microscope was designed for the visualization of monolayers. The presence of a surfactant monolayer at an interface introduces a refractive index change, and therefore a change in reflectivity. *See* Reference [71].
Bridging Flocculation	→ Polymer Bridging.
Brighteners	→ Optical Brighteners.
Brightening Agents	→ Optical Brighteners.
Bright-Field Illumination	A kind of illumination for microscopy, in which the illumination of a specimen is arranged so that transmitted light remains in the optical path of the microscope and is used to form the magnified image. This illumination differs from the arrangement in Dark-Field Illumination.
Brookfield Viscometer	A commercial instrument brand-name that has become a general use term; better known than the Saybolt type. The basic model is a direct torque-reading, mechanically simple viscometer of the concentric-cylinder type, which is used under laboratory and field conditions for determining the viscosities of materials over a (usually limited) range of shear rates. Similar to the Fann Viscometer (another commercial brand).
Brown, Robert (1773–1858)	Although primarily a botanist, the Scottish Brown is known to colloid science for his 1827 discovery that dispersed particles in water move about randomly, even when the water itself appears motionless. The phenomenon, explained later by others, is due to bombardments of the particles by water molecules and is known as Brownian Motion. *See* Reference [72].
Brownian Motion	Random fluctuations in the density of molecules at any location in a liquid, due to thermal energy, cause other molecules and small dispersed particles to move along random pathways. The random particle motions are termed Brownian Motion and are most noticeable for particles smaller than a few micrometres in diameter.
Brownstock	In the kraft chemical pulping of cellulose wood fibres (for paper making) alkaline solutions are used to degrade and dissolve lignin and allow the cellulose fibres to be separated, washed, and possibly bleached. After kraft cooking, the degradation products, called kraft lignin, are dissolved and dispersed in the liquor, adsorbed on the cellulose, and trapped in fibre cell walls giving them a brown colour, hence the term brownstock. Brownstock washing involves separating as much lignin as possible from the pulp.

Bruggeman Equation	An equation for predicting the conductivities or relative permittivities of dispersions. *See* Tables 9 and 10.
Brunauer, Stephen (1903–1986)	A Hungarian-born American chemist known for his work on the adsorption of gases by solids, including the transition from monolayer to multilayer adsorption. He also contributed to the areas of chemisorption and explosives. He is known in interface science for the BET (Brunauer-Emmett-Teller) theory of multilayer adsorption. *See* References [73, 74].
Brunauer-Emmett-Teller Isotherm	(BET Isotherm) An adsorption isotherm equation that accounts for the possibility of multilayer adsorption and different enthalpy of adsorption between the first and subsequent layers. Five "types" of adsorption isotherm are usually distinguished. These types are denoted by roman numerals and refer to different characteristic shapes. → Adsorption Isotherm.
BS&W	→ Basic Sediment and Water.
Bubble Point	The gas pressure at which gas bubbles are generated and evolved from a liquid.
Bubble	A gas globule surrounded by liquid, either bulk liquid (as in the bubbles rising through boiling water) or a liquid film (as in the common soap bubbles).
Bubble Snap-Off	→ Snap-Off.
Buckminsterfullerene	→ Fullerene.
Buckyball	→ Fullerene.
Buckytube	→ Carbon Nanotube, → Fullerene.
Builder	A chemical compound added into detergent formulations to aid oil emulsification (by raising pH and to complex and solubilize hardness ions). Example: sodium tripolyphosphate.
Bulk Foam	Any foam for which the length scale of the confining space is greater than the length scale of the foam bubbles. The converse case categorizes some foams in porous media, distinguished by the term "lamellar foam". → Foam, → Foam Texture.
Bulk Nanoparticles	Nanoparticles that have been produced by an industrial-scale process. Examples: carbon black, titanium dioxide, and fumed silica.
Bulking Agent	A material that is added to a formulation that increases the quantity of formulation required for a process without actually changing the formulation's reactivity. Example: Barium sulfate is sometimes added during processing to increase the density of polyurethane (solid) foam.

Bunsen, Robert (Wilhelm) (1811–1899)	A German chemist and pioneer of chemical spectroscopy. He and Kirchhoff discovered the use of spectroscopy in chemical analysis and developed the spectroscope. They discovered the elements caesium and rubidium. Bunsen invented a number of laboratory instruments and is known for his studies of the spectra (colours) produced by different materials when heated to incandescence in a flame. Eponyms include the Bunsen (electrochemical) cell, Bunsen ice calorimeter, and the Bunsen burner.
Butterfat Spread	→ Spreadable Fats.
Butter Mixture	→ Spreadable Fats.

C

Cabannes Factor	A factor used in light-scattering analysis to correct for particle anisotropy. $\rightarrow$ Depolarization.
C-AFM	Contact-Mode Atomic Force Microscopy. $\rightarrow$ Friction Force Microscopy.
Calcination	Producing, or modifying, a powder material by heating it to a high temperature in a dry atmosphere.
Calculation of Phase Inversion in Concentrated Emulsions	(CAPICO) A system in which potential cosmetic emulsion ingredients are numerically categorized so that one may calculate their influence on the phase inversion temperature of a formulated emulsion. *See* Reference [75].
Canal Viscometer	An instrument used to measure interfacial viscosity by measuring the flow rate of a surface or interfacial layer through a narrow channel or canal. The analysis is essentially a two-dimensional analog of the capillary viscosity method for fluids.
Cantilever	$\rightarrow$ Microcantilever.
Capacitance of the Electric Double Layer	The integral capacitance of the electric double layer (per unit area) is the charge density at the outer Helmholtz plane divided by the electric potential at the outer Helmholtz plane. The differential capacitance of the electric double layer (per unit area) is the partial derivative of the charge density with respect to the potential at the outer Helmholtz plane.
CAPICO	$\rightarrow$ Calculation of Phase Inversion in Concentrated Emulsions.
Capillarity	A general term referring either to the general subject of, or to the various phenomena attributable to the forces of surface or interfacial tension. The Young-Laplace equation is sometimes referred to as the equation of capillarity.

Dictionary of Nanotechnology, Colloid and Interface Science. Laurier L. Schramm
Copyright © 2008 WILEY-VCH Verlag GmbH & Co. KGaA, Weinheim
ISBN: 978-3-527-32203-9

Capillary	A tube with very small internal diameter. Originally the term referred to cylindrical tubes whose internal diameters were of similar dimension to hairs.
Capillary Condensation	The process by which multilayer adsorption from a vapour into a porous medium proceeds to the point at which pore spaces become filled with condensed liquid from the vapour. The dimensions of the pore must be large enough that the concept of a separating meniscus retains a physical meaning.
Capillary Constant	For two phases in contact, the capillary constant, a, is given by $a^2 = 2\gamma/\Delta\rho g$; that is, the square of the capillary constant equals the ratio of twice the surface (or interfacial) tension to the product of gravitational constant and the density difference between the phases. This dimensionless number is used in considerations of capillarity, such as in capillary rise.
Capillary Electrometer	An instrument used to determine electrocapillary curves. A column of mercury is attached to an electrochemical cell that is used to apply an electric potential to the mercury-aqueous solution interface. The interfacial tensions corresponding to different states of applied electric potential were originally determined by capillary rise, but subsequent capillary electrometers have used other interfacial tension methods. $\rightarrow$ Electrocapillarity.
Capillary Electrophoresis	An electrophoresis technique in which dissolved molecules are separated, on the basis of charge, in a fused silica capillary tube. Also termed "capillary zone electrophoresis". $\rightarrow$ Micellar Electrokinetic Capillary Chromatography, $\rightarrow$ Isotachophoresis, $\rightarrow$ Isoelectric Focussing.
Capillary Flow	Liquid flow in response to a difference in pressures across curved interfaces. $\rightarrow$ Capillary Pressure.
Capillary Forces	The interfacial forces acting among oil, water, and solid in a capillary or in a porous medium. These forces determine the pressure difference (capillary pressure) across an oil-water interface in the capillary or in a pore. Capillary forces are largely responsible for oil entrapment under typical petroleum reservoir conditions.
Capillary Number	(N_c) A dimensionless ratio of viscous to capillary forces. One form gives N_c as Darcy velocity times viscosity of displacing phase divided by interfacial tension. It is used to provide a measure of the magnitude of forces that trap residual oil in a porous medium.
Capillary Pressure	(1) The pressure difference across an interface between two phases. When the interface is contained in a capillary, it is sometimes referred to as the "suction pressure".

(2) In petroleum reservoirs it is the local pressure difference across the oil-water interface in a pore contained in a porous medium. One of the liquids usually preferentially wets the solid, and therefore the capillary pressure is normally taken as the pressure in the nonwetting fluid minus that in the wetting fluid.

Capillary Ripples

Surface or interfacial waves caused by perturbations of an interface. Where the perturbations are caused by mechanical means (e.g., barrier motion) the transverse waves are known as "capillary ripples" or "Laplace waves", and the longitudinal waves are known as "Marangoni" waves. The characteristics of these waves depend on the surface tension and the surface elasticity. This feature forms the basis for the capillary wave method of determining surface or interfacial tension.

Capillary Rise

The tendency, and process, for a liquid to rise in a capillary. Example: Water rises in a partially immersed glass capillary. Negative capillary rise occurs when the liquid level in the capillary falls below the level of bulk liquid, as when a glass capillary is partially immersed in mercury. Capillary rise forms the basis for a method of determination of surface or interfacial tension.

Capillary Viscometer

An instrument used for the measurement of viscosity in which the rate of flow through a capillary under constant applied pressure difference is determined. This method is most suited to the determination of Newtonian viscosities. There are various designs, among which are the Ostwald and Ubbelohde types.

Capillary Wave Method

→ Capillary Ripples.

Capillary Waves

→ Capillary Ripples.

Capillary Zone Electrophoresis

→ Capillary Electrophoresis.

Carbon Aerogel

A carbon nanofoam. → Nanofoam.

Carbon Black

Amorphous (black) carbon particles created by the decomposition of hydrocarbons; usually thermal decomposition. The particles are usually of colloidal, possibly nanoscale, size. Such particles exhibit large specific surface areas and are capable of acting as an effective adsorbent for some substances.When produced by the decomposition, or partial combustion, of natural gas it is termed gas black or channel black. Carbon black is used as a pigment in inks, lubricants, and rubber. → Lampblack.

Carbon Nanofoam

→ Nanofoam.

Carbon Multi-wall Nanotube

(CMWNT) → Carbon Nanotube.

Carbon Nanotube

(1) (CN or CNT) A self-organized carbon nanostructure in the form of a tube. A single-wall carbon nanotube is composed of a single sheet of graphite, termed graphene, rolled-up into a tube. Carbon nanotubes are usually made by vaporizing carbon electrodes in an arc-discharge, by laser ablation, or by chemical vapor deposition. Typically, carbon nanotubes are 1 to $10 \mu m$ in length, although lengths up to about $200 \mu m$ have been made. Carbon single-wall nanotubes (SWNT) can have metallic or semiconducting properties. Example: single-walled carbon nanotubes have been used in the preparation of linear-$(C_{60}O)_n$ polymers [76]. Also termed bucky-tubes, $\rightarrow$ Iijima, Sumio.

(2) Carbon multi-wall nanotubes (MWNT) comprise several concentric single-wall nanotubes and are typically 10 to $40 nm$ in diameter. Most multi-wall carbon nanotubes have metallic properties. Multi-wall nanotubes can be: concentric (c-MWNT), herringbone (h-MWNT, in which the graphenes make an angle with respect to the nanotube axis), or either of these with an additional "bamboo" texture (bc-MWNT or bh-MWNT, in which some graphenes are oriented perpendicular to the nanotube axis). (Filaments with the bamboo texture are not open along their full length and may be termed nanofibres.)

Carbon Single-Wall Nanotube

(CSWNT) $\rightarrow$ Carbon Nanotube.

Carrageenan

A water-soluble mixture of sulfated linear polysaccharides derived from seaweeds such as Irish Moss. Carrageenan is considered to have a number of different structural types that are designated by different Greek letter prefixes, for example κ-carrageenan. Carrageenan sols can be quite viscous and readily form gels, and they can be used to stabilize certain suspensions, foams, and emulsions. Carrageenan is used in many different applications including food processing. *See also* Reference [35], $\rightarrow$ Seaweed Colloids, $\rightarrow$ Hydrocolloid.

Carreau Equation

An empirical equation used to describe the non-Newtonian flow behaviour of polymer solutions. *See* Reference [77] and Table 11.

Carrier Flotation

A variation on standard froth flotation in which small-sized particles become attached to the surfaces of larger (carrier) particles. The carrier particles attach to the air bubbles and the combined aggregates of small desired particles, carrier particles, and air bubbles float to form the froth. Example: using limestone particles as carriers in the flotation removal of fine iron and titanium oxide mineral impurities from kaolinite clays. $\rightarrow$ Emulsion Flotation, $\rightarrow$ Floc Flotation, $\rightarrow$ Roughing Flotation, $\rightarrow$ Scalping Flotation, $\rightarrow$ Scavenging Flotation, $\rightarrow$ Froth Flotation.

Cascade Impactor	$\rightarrow$ Impactor.
Cascade Mixing	In electron and other surface spectroscopies, the rearrangement of atoms in a solid caused by collisions between incident particles and the atoms, within the penetration depth of the incident particles.
Casein	A surface-active class of protein found in milk, casein is the main protein (as calcium caseinate) in milk (in this case casein is sometimes termed caseinogen). There are four types of casein molecules of which most complex with calcium phosphate and reside in the interior of the casein micelles. One type, κ-casein, forms stabilizing surface layers. In the making of cheese casein is precipitated by rennet (an enzyme) and flocculated to form curds (in this case the casein is sometimes termed paracasein). A range of types of food emulsions are stabilized by κ-casein, casein micelles, and/or fragments of casein micelles. Caseinate salts are also used as surfactants in the manufacture of many other products. $\rightarrow$ Curds, $\rightarrow$ Whey.
Caseinogen	$\rightarrow$ Casein.
Casimir Effect	$\rightarrow$ Casimir Forces.
Casimir Forces	Interaction forces, that are net attractive, between two facing mirror plates, due to fluctuations in their surrounding electromagnetic fields. The net attractive force is proportional to cross-sectional area of the plates and inversely proportional to the sixth power of the separation distance between the plates. The Casimir (attractive) effect can influence the operation of micro- and nano-mechanical devices.
Casson Equation	A semi-empirical model equation used to describe flow behaviour. It allows one to obtain the shear stress for pure matrix liquid by subtracting the yield stress (the Casson yield point) from the total stress, at a given shear rate. Example: used in the evaluation of chocolate. For this and other models *see* Reference [21].
Casson Yield Point	$\rightarrow$ Casson Equation.
Catalyst	A substance that increases the rate or yield of a reaction. Heterogeneous catalysis refers to the situation in which the catalytic reactions occur at a surface or interface between two phases. In practice heterogeneous catalysts tend to have high specific surface areas. Homogeneous catalysis refers to the situation in which the reaction(s) take place within a single phase.
Catalyst Poison	$\rightarrow$ Poison.
Catalyst Promoter	$\rightarrow$ Promoter.

Cataphoresis	$\rightarrow$ Electrophoresis.
Cathode	An electrode at which a net negative current flows. The predominant chemical reaction here is reduction.
Cation-Exchange Capacity	The capacity for a substrate to adsorb hydrated cationic species while simultaneously desorbing (exchanging) an equivalent charge quantity of other cationic species. Example: This property is used to characterize clay minerals that can have very large cation-exchange capacities and also significant anion-exchange capacities. $\rightarrow$ Ion Exchange.
Cationic Surfactant	A surfactant molecule that can dissociate to yield a surfactant ion whose polar group is positively charged. Example: cetyltrimethylammonium bromide, $CH_3(CH_2)_{15}N^+(CH_3)_3Br^-$.
Cavitation	The spontaneous formation of bubbles in a liquid when subjected to a pressure that falls below a critical value. Such low pressures can be produced where a liquid is being accelerated to high velocities, such as in propellers, turbines and pumps. In some cases the bubbles produced will rapidly collapse, producing shock waves.
CCC	$\rightarrow$ Critical Coagulation Concentration.
CCT	$\rightarrow$ Critical Coagulation Temperature.
CEA	$\rightarrow$ Colloidal Emulsion Aphrons.
Cell Membrane	Thin films composed of lipids and proteins that cover the surfaces of cells. Also termed "plasma membranes" or "plasmalemma".
CELS	$\rightarrow$ Characteristic Energy-Loss Spectroscopy.
Celsius, Anders (1701–1744)	A Swedish astronomer famous for his astronomical catalogues and his participation in the "Lapland expedition" which confirmed Newton's opinion that the shape of the earth is an ellipsoid flattened at the poles. He carried out many geographic and meteorological observations, for which he constructed the famous Celsius thermometer (sometimes termed centigrade), giving the boiling point of water 0 degrees Celsius, and the freezing point 100 degrees Celsius.
Cementation	The process by which a material is precipitated at the meeting point of multiple particles, binding them together.
Centrifugal Separator	$\rightarrow$ Separator.
Centrifuge	An apparatus in which an applied centrifugal force is used to achieve a phase separation by sedimentation or creaming. For centrifuges operating at very high relative centrifugal forces (so-called g-forces) the terms supercentrifuge (ca. tens of thousands RCF or gs) or ultracentrifuge (ca. hundreds of thousands RCF or gs) are used. $\rightarrow$ Relative Centrifugal Force.

Centrifuge Test	$\rightarrow$ Bottle Test.
CFC	Critical flocculation concentration. $\rightarrow$ Critical Coagulation Concentration.
CFT	Critical flocculation temperature. $\rightarrow$ Critical Coagulation Temperature.
Channel Black	$\rightarrow$ Carbon Black.
Chapman, David Leonard (1869–1958)	A British chemist who worked extensively in the area of gas reactions and their kinetics. He is remembered in colloid science for his theory of the diffuse double layer at a charged interface, as embodied in the Gouy-Chapman and Gouy-Chapman-Stern theories. Other eponyms include the Chapman equation (gaseous explosion rates). *See* References [78, 79].
Characteristic Adsorption Curve	$\rightarrow$ Characteristic Isotherm.
Characteristic Energy-Loss Spectroscopy	(CELS) A technique for studying surface composition and surface energy states. Inelastically scattered electrons, having lower energy than the incident beam, are used to form the image pattern, and the characteristic energy losses of the scattered electrons are determined. Similar techniques include electron loss spectroscopy (ELS), which is also termed "electron energy loss spectroscopy (EELS)", "electron impact spectroscopy (EIS)", and "high-resolution electron energy loss spectroscopy (HREELS)".
Characteristic Isotherm	An adsorption isotherm involving multilayer adsorption in which the equilibrium quantity of a material adsorbed is essentially related to the composition of the bulk phase, at constant temperature, by a single relationship for a given adsorbate independent of the nature of the adsorbent. Also termed the "characteristic adsorption curve". $\rightarrow$ Adsorption Isotherm.
Characteristic X-Rays	Photons, emitted by ionized atoms, that have a particular distribution of intensity and energy that is characteristic of the atomic number and chemical environment of those atoms. In XPS characteristic X-rays refer to the spectrometer source.
Charged Agar	A combination of two of the types of polysaccharide structure that constitute agar, pyruvic acid acetal, and sulfated galactan. Charged agar is also termed "agaropectin". $\rightarrow$ Agar.
Charge Density	In colloidal systems, the quantity of charge at an interface, expressed per unit area.
Charge-Determining Ions	$\rightarrow$ Potential-Determining Ions.
Charge of the Micelle	$\rightarrow$ Micellar Charge.

Charge Reversal	The process wherein a charged substance is caused to take on a new charge of the opposite sign. Such a change can be brought about by any of oxidation, reduction, dissociation, ion exchange, or adsorption. Example: The adsorption of cationic polymer molecules onto negatively charged clay particles can exceed the requirements for charge neutralization and thus cause charge reversal.
Charging Potential	In surface analysis the charging potential is the electrical potential caused by irradiation of the surface of an insulating target. Heterogeneous target surfaces may exhibit a distribution of charging potentials.
Chemical Adsorption	→ Chemisorption.
Chemical Disassembler	→ Disassembler.
Chemical Nanosensor	→ Nanosensor.
Chemical Vapour Deposition	(CVD) A gas-phase particle synthesis method in which one or more of the feedstock materials goes through a gas phase and then reacts with or decomposes on the target surface. Example: CVD is used to produce thin semiconductor films and has been used to make carbon nanotubes. → Physical Vapour Deposition.
Chemical Vapour Growth	→ Chemical Vapour Synthesis.
Chemical Vapour Synthesis	A method of producing particles in which a chemical reaction, such as oxidation, reduction, or pyrolysis, produces a vapour that subsequently condenses. Also termed chemical vapour growth. Example: this is one of the methods used to synthesize carbon nanotubes.
Chemisorption	(Chemical Adsorption) The adsorption forces are of the same kind as those involved in the formation of chemical bonds. The term is used to distinguish chemical adsorption from physical adsorption, or physisorption, in which the forces involved are of the London-van der Waals type. Some guidelines for distinguishing between chemisorption and physisorption are given by IUPAC in Reference [4].
Chemosynthesis	→ Nanotechnology.
Chinese Ink	→ India Ink.
Chi Potential	→ Jump Potential.
Chitin	A mucopolysaccharide, soluble in organic liquids, derived from various plants, fungi, and some marine animals, such as shellfish. Derivatized (deacetylated) chitin is referred to as chitosan. Chitin and chitosan solutions can be quite viscous, depending on the solvent, usually an organic liquid or an acid solution. They are used as coagulating agents in food industry applications. *See* Reference [35], → Marine Colloids.

Chitosan	Derivatized (deacetylated) chitin. $\rightarrow$ Chitin.
Chocolate	A food colloid comprising a suspension of cocoa, sugar and solid milk particles in a continuous fat phase.
Chocolate Mousse Emulsion	A name frequently used to refer to the W/O emulsions of high water content that are formed when crude oils are spilled on the oceans. The name reflects the color and very viscous consistency of these emulsions. It has also been applied to other petroleum emulsions of similar appearance.
Chromatography	A process or procedure in which flow through a permeable porous medium or through a capillary causes components in a mixture to become separated as a result of their different affinities for the mobile and stationary phases. Several kinds of liquid chromatography are capable of differentiating large molecules on the basis of size: In gel filtration (gel permeation) chromatography the stationary phase is a polymer gel or porous bead packing capable of sorbing smaller size molecules while larger size molecules pass through. The term can also refer to the separation of subsurface contaminant plumes in the environment. $\rightarrow$ Hydrodynamic Chromatography.
CLA	$\rightarrow$ Colloidal Liquid Aphrons.
Classifier	A machine used to separate particles of specified size ranges. Wet classifiers include settling tanks, centrifuges, hydrocyclones, and vibrating screens. Dry classifiers, also termed "air classifiers", use gravity or centrifugal settling in gas streams. *See also* Reference [80].
Classifier Mill	A kind of mechanical impact mill or jet mill for size reduction (comminution) that also incorporates a particle classifier.
Clausius-Mossotti Factor	An equation reflecting differences in permittivity between a colloidal species (ε_c) and the surrounding medium (ε_m) as $\kappa = (\varepsilon_c - \varepsilon_m)/(\varepsilon_c + 2\varepsilon_m)$. K, the Clausius-Mossotti factor, is used in dielectrophoresis. If $K > 0$, the motion of a dipolar colloidal species is toward the most intense region of an imposed, nonhomogeneous electric field. If $K < 0$ the species moves toward the least intense region, and for $K = 0$ there is no dielectrophoretic motion. $\rightarrow$ Dielectrophoresis, $\rightarrow$ Levitator.
Clays	(1) The term "clay minerals" refers to the aluminosilicate minerals having two- or three-layer crystal structure. These minerals typically exhibit high specific surface area, significant surface charge density (cation-exchange capacity), and low hydraulic conductivity. Examples: montmorillonite, kaolinite, illite. (2) The term "clays" is sometimes used to distinguish particles having sizes of less than about 2 to 4 μm, depending upon the size classification system used. In this sense the term includes any suitably fine-grained solids, including nonclay minerals. *See* Table 12.

Closed-Cell Foam	A solid foam in which individual gas bubbles are completely separated from their neighbours so that gas exchange can take place only by diffusion through the walls. Rigid solid foams are often formulated to be closed-cell foams for maximum thermal insulating properties. Flexible solid foams are often formulated to be open-cell foams for maximum flexibility and breathability. *See* Reference [81].
Clotted Soap	$\rightarrow$ Middle Soap.
Cloud	An aerosol of liquid droplets (as in rain cloud) or an aerosol of solid particles (such as dust cloud or smoke cloud). $\rightarrow$ Aerosol of Liquid Droplets, *see* Table 5.
Cloud Point	(1) Aqueous Solutions: The transition temperature above which a nonionic surfactant or wax loses some of its water solubility and becomes ineffective as a surfactant. The originally transparent surfactant solution becomes cloudy because of the separation of a surfactant-rich phase. Cloud points are typically reported on the basis of tests for a specified surfactant concentration such as 1 mass%. $\rightarrow$ Coacervation. (2) Petroleum Liquids: The temperature at which waxes or other substances begin to separate from solution.
Cloud Point Extraction	(CPE) An extraction/separation technique in which a desired analyte species is trapped in the micelles of an added surfactant (usually a nonionic surfactant). A temperature change is then induced to move the system past the cloud point of the surfactant, causing a phase separation of surfactant plus analyte. Also termed micelle-mediated extraction.
CLS	$\rightarrow$ Light Scattering
Clustering	In food processing a kind of oil droplet aggregation termed "clustering" is used to increase viscosity and partially gel products such as cream. Clustering is achieved when emulsion droplets are subdivided to the point where the total interfacial area in the emulsion is greater than the surface-covering capability of the adsorbing proteins. This leads to proteins bridging between pairs of droplets (i.e., bridging flocculation).
CMC	$\rightarrow$ Critical Micelle Concentration.
CMT	$\rightarrow$ Critical Micelle Temperature.
c-MWNT	Concentric Multi-Wall Nanotube. $\rightarrow$ Carbon Nanotube.
CMWNT	Carbon Multi-Wall Nanotube. $\rightarrow$ Carbon Nanotube.
CN or CNT	$\rightarrow$ Carbon Nanotube.
Coacervate	$\rightarrow$ Coacervation.

Coacervation	When a lyophilic colloid loses stability, sometimes due to the addition of a new component, a separation into two liquid phases may occur. This process is termed "coacervation". The phase that is more concentrated in the colloid is the coacervate, and the other phase is the equilibrium solution. $\rightarrow$ Cloud Point.
Coactive Surfactant	The secondary surfactant(s) in a detergent formulation. $\rightarrow$ Detergent.
Coadsorption	The adsorption of more than one species simultaneously.
Coagulation	$\rightarrow$ Aggregation.
Coagulation Half-Life	$\rightarrow$ Aggregation Time.
Coagulation Time	$\rightarrow$ Aggregation Time.
Coagulum	The dense aggregates formed in coagulation are referred to, after separation, as coagulum. $\rightarrow$ Aggregation.
Coalescence	The merging of two or more dispersed species into a single one. Coalescence reduces the total number of dispersed species and also the total interfacial area between phases. In emulsions and foams coalescence can lead to the separation of a macrophase, in which case the emulsion or foam is said to break. The coalescence of solid particles is termed "sintering".
Coalescor	An emulsion separator that functions based on differences in interfacial tension and wetting in a porous medium. See Reference [37].
Coarse Sand	$\rightarrow$ Sand, see Table 12.
Coaxial Nanowire	$\rightarrow$ Nanowire.
Coefficient of Friction	$\rightarrow$ Friction.
Cohesion	The tendency of a body of a substance to resist being mechanically pulled apart.
Cohesive Energy Density	$\rightarrow$ Solubility Parameter.
Coincidence	In particle characterization, the detection of two or more particles as if they were a single particle.
Co-Ions	In systems containing large ionic species (colloidal ions, membrane surfaces, etc.), co-ions are those that, compared with the large ions, have low molecular mass and the same charge sign. For example, in a suspension of negatively charged clay particles containing dissolved sodium chloride, the chloride ions are co-ions and the sodium ions are counterions. $\rightarrow$ Counterions.
Collagen	Insoluble protein from tendons and inner layers of skin. Glue and gelatin can be produced from collagen. $\rightarrow$ Keratin.

Collapse Pressure	The film pressure required to cause a surface or interfacial mono-molecular film to compress to an area that will no longer support a monolayer of adsorbed species; thus it will distort and collapse.
Collector	(1) Flotation: A surfactant used in froth flotation to adsorb onto solid particles, make them hydrophobic, and thus facilitate their attachment to gas bubbles. $\rightarrow$ Froth Flotation; $\rightarrow$ Modifier; $\rightarrow$ Frothing Agent. (2) Particle Separation: A plate positioned in the gas flow path, in an impactor, and used to collect larger-sized particles. $\rightarrow$ Impactor.
Colligative Properties	Properties of matter that depend upon the number of species rather than upon their mass or activity.
Colloid	In the early 1800s Thomas Graham studied the diffusion, osmotic pressure, and dialysis properties of a number of substances, including a variety of solutes dissolved in water [1–3]. Some substances diffused quite quickly through parchment paper and animal membranes and formed crystals when dried. Other substances diffused very slowly if at all through the parchment or membranes and apparently did not form crystals when dried. Graham proposed that the former group of substances, which included simple salts, be termed "crystalloids" and the latter group, which included albumen and gums, be termed "colloids". The test of crystal formation later turned out to be too restrictive; the distinction of crystalloids versus colloids was dropped, and the noun "colloid" was eventually replaced by the adjective Colloidal. $\rightarrow$ Colloidal Dispersion. Prior to Graham's time such finely divided substances held in suspension were referred to as "being in pseudo-solution". *See* Reference [39].
Colloidal	A state of subdivision in which the particles, droplets, or bubbles dispersed in another phase have at least one dimension between approximately 1 and 1000 nm. In some literature distinctions are drawn among colloids according to the number of their dimensions that fall into the colloidal range: laminar colloids (e.g., a thin film), fibrillar colloids (e.g., fibres) and corpuscular colloids (e.g., suspensions and emulsions). $\rightarrow$ Colloidal Dispersion, $\rightarrow$ Micro-technology, $\rightarrow$ Nanotechnology, $\rightarrow$ Picotechnology.
Colloidal Dispersion	A system in which colloidal species are dispersed in a continuous phase of different composition or state. *See* Table 1.
Colloidal Electrolyte	An electrolyte that dissociates to yield ions at least one of which is of colloidal or near-colloidal size. Example: ionic surfactant micelles.
Colloidal Emulsion Aphrons	(CEA) $\rightarrow$ Colloidal Liquid Aphrons.

Colloidal Gas Aphrons	→ Microgas Emulsions.
Colloidal Liquid Aphrons	(CLA) A kind of emulsion in which micrometre-size dispersed droplets have an unusually thick stabilizing film and exist clustered together as opposed to separated, nearly spherical droplets. The stabilizing aqueous film, sometimes called a "soapy shell," is thought to have inner and outer surfactant monolayers. Also termed "aphrons" or "colloidal emulsion aphrons (CEA)". → Microgas Emulsions.
Colloidal Production Methods	Methods in which aqueous solutions are mixed, under controlled conditions of temperature and pressure, to precipitate suspended nanoparticles and/or colloidal particles. In some cases the particles are then recovered by filtering or spray drying.
Colloidal Processing	In ceramics, a variation of slip-casting in which a stabilized colloidal dispersion of particles is poured into a mold for sintering.
Colloid Anion	A colloidal species with a net negative electric charge.
Colloid Cation	A colloidal species with a net positive electric charge.
Colloid Mill	A high-shear mixing device used to prepare colloidal dispersions of particles or droplets by size reduction (comminution). Also termed "dispersion mill".
Colloid Osmotic Pressure	When a colloidal system is separated from its equilibrium liquid by a semipermeable membrane, not permeable to the colloidal species, the colloid osmotic pressure is the pressure difference required to prevent transfer of the dissolved, noncolloidal species. Also referred to as the "Donnan pressure". The reduced osmotic pressure is the colloid osmotic pressure divided by the concentration of the colloidal species. → Osmotic Pressure.
Colloidosome	Encapsulated droplets of oil or water having a shell of close-packed, bound particles. The pores between the encapsulating particles control the permeability between external and internal phases. In some cases the particles are sintered together. The term colloidosome is by analogy with liposome. → Nanocapsule, → Vesicle. See Reference [82].
Colloid Science	The study of materials and phenomena at the colloidal scale (typically 1 to 1,000 nm). This field overlaps significantly with the field of nanoscience. → Nanoscience, → Nanotechnology.
Colloid Stability	In colloid science the term "colloid stability" means that a specified process that causes the colloid to become a macrophase, such as aggregation, does not proceed at a significant rate. Colloid stability is different from thermodynamic stability [4]. The term colloid stability must be used with reference to a specific and clearly

defined process, for example, a colloidally metastable emulsion may signify a system in which the droplets do not participate in aggregation, coalescence, or creaming at a significant rate. → Kinetic Stability, → Thermodynamic Stability.

Colloid Titration
A method for the determination of charge, and the zero point of charge, of colloidal species. The colloid is subjected to a potentiometric titration with acid or base to determine the amounts of acid or base needed to establish equilibrium with various pH values. By titrating the colloid in different, known concentrations of indifferent electrolyte, the point of zero charge can be determined as the pH for which all the isotherms intersect. → Point of Zero Charge.

Colloid Vibration Current
(CVC) → Ultrasound Vibration Potential.

Colloid Vibration Potential
(CVP) → Ultrasound Vibration Potential.

Columnar Thin Film
(CTF) → Sculptured Thin Film.

Comminution
The reduction of particles, or other dispersed species, into smaller sizes, by fracture caused by friction and wear. Also called attrition. Examples of comminution machines include agitator ball mills, colloid mills, cutting mills, disk mills, jet mills, mechanical impact mills, ring-roller mills, and roll crushers. *See* Reference [63], → Ablation.

Common Black Film
→ Black Film.

Compaction
→ Subsidence.

Complex Coacervation
The process of coacervation when caused by the interaction of oppositely charged colloids.

Composite
A material that contains particles of a second substance introduced to increase material strength.

Composite Clay Nanostructures
→ Pillar Interlayered Clay Minerals.

Composite Isotherm
A (no longer recommended) term referring to the Surface Excess Isotherm.

Compressibility
(1) The ratio of relative volume change to applied compressional stress.
(2) For film compressibility, → Film Elasticity.

Compressional Modulus
→ Film Elasticity.

Concentric Cylinder Rheometer
→ Rheometer.

Concentric Multi-Wall Nanotube	(c-MWNT) → Carbon Nanotube.
Condensate	Any light hydrocarbon liquid mixture obtained from the condensation of hydrocarbon gases. Condensate typically contains mostly propane, butane, and pentane.
Condensation Methods	The class of methods used for preparing colloidal dispersions in which precipitation from either solution or chemical reaction is used to create colloidal species. The colloidal species are built up by deposition on nuclei that may be of the same or different chemical species. If the nuclei are of the same chemical species, the process is referred to as homogeneous nucleation; if the nuclei are of different chemical species, the process is referred to as heterogeneous nucleation. → Dispersion Methods.
Condensation Particle Counter	(CPC) An instrument used to count ultrafine particles (nanoparticles) in aerosols of solid particles. A CPC operates by condensing a vapor onto a sample of nanoparticles in order that they grow to sizes that can be detected by a conventional optical counting technique.
Condensed State	In adsorption, the state of a monolayer when it can be considered to behave like a layer of liquid (close-packed molecules).
Conductivity of a Dispersion	*See* Table 10.
Cone-Cone Rheometer	→ Rheometer.
Cone-Plate Rheometer	→ Rheometer.
Confocal Microscopy	A microscopic technique used to produce three-dimensional images of specimens that actually have considerable depth. A series of shallow depth-of-field image slices through a thick specimen are obtained. The three-dimensional image is then obtained by reconstruction so that no out-of-focus elements contribute to the final image.
Consistency	An empirical or qualitative term referring to the relative ease with which a material can be deformed or made to flow. It is a reflection of the cohesive and adhesive forces in a mixture or dispersion. → Atterberg Limits.
Consolute Temperature	The temperature at which two partially miscible liquids become fully miscible. Stated differently, this is the upper temperature for which two liquid components remain immiscible, and in this sense is sometimes termed the upper consolute temperature or the upper critical solution temperature.
Contact Angle	When two immiscible fluids (e.g., liquid-gas or oil-water) are both in contact with a solid, the angle formed between the solid surface and the tangent to the fluid-fluid interface intersecting the three-

phase contact point is termed the "contact angle". It is essential to state through which phase the contact angle is measured. By convention, if one of the fluids is water then the contact angle is measured through the water phase; otherwise, the contact angle is usually measured through the most dense phase. Distinctions may be made among advancing, receding, or equilibrium contact angles. Contact angles are important in areas such as liquid wetting, imbibition, and drainage.

Contact-Angle Hysteresis	A phenomenon manifested by differing values of advancing and receding contact angles in the same three-phase contact system. Both may differ from the equilibrium contact angle. → Contact Angle, → Wetting Hysteresis.
Contact-Mode Atomic Force Microscopy	(C-AFM) → Friction Force Microscopy.
Continuous Phase	In a colloidal dispersion, the phase in which another phase of particles, droplets, or bubbles is dispersed. Sometimes referred to as the external phase. Continuous phase is the opposite of dispersed phase. → Dispersed Phase.
Copolymer	A polymer composed of more than one kind of monomer. → Block Copolymer.
Corpuscular Colloid	→ Colloidal.
Coriolis Effect	An apparent inertial force that develops when an object of finite mass moves radially along a rotational path. The object appears to follow a deviating path although it is actually the coordinate system that is moving. From the perspective of the rotating object's frame of reference the apparent force experienced is proportional to the mass, linear velocity and angular velocity. → Corolis, (Gaspard) Gustav de, → Ekman Effect.
Coriolis Flow Meter	A mass flow meter used for direct, real-time measurements of mass, flow rate, and/or density. In industrial settings a Coriolis meter is usually a tube, or a pair of tubes, that are made to vibrate. With a fluid flowing through the tube(s), Coriolis forces will cause the tube(s) to twist somewhat. The angle of twist increases with flow rate. By making the tube vibrate at its natural frequency, changes in fluid density are measured as changes in the frequency of vibration. Also known as an inertial flow meter, the Coriolis flow meter can be applied to flowing foams, suspensions and aerosols.
Coriolis Force	→ Coriolis, (Gaspard) Gustav de, → Coriolis Effect.
Coriolis, (Gaspard) Gustav de (1792–1843)	A French physicist known for his work in mechanics and dynamics, and for his discovery of the Coriolis inertial force (in 1832) that acts on rotating surfaces, at right angles to their direction of

rotation. Such effects are important, for example, in oceanography, ballistics, and meteorology. Coriolis established the use of the word work (force/distance) in mechanics, and introduced the term kinetic energy. Additional eponyms include the Coriolis flow meter.

Cosmeceuticals   → Cosmoceuticals.

Cosmoceuticals   Personal care products, particularly cosmetics, that are applied topically and that provide pharmaceutical benefits in addition to their cleansing, protecting, and/or perfuming actions. Example: the first cosmoceutical, developed in 1984, was a cosmetic cream that contained alpha hydroxy acids. Sometimes termed Cosmeceuticals.

Cosorption Lines   Contours of equal surface activity, as measured by the Gibbs surface excess concentrations, plotted on phase diagrams. *See* Reference [8], p 131.

Cosurfactant   A surfactant that may be added to a system to enhance the effectiveness of another surfactant. The term "cosurfactant" has also been improperly used to describe non-surface- active species that enhance a surfactant's effectiveness, such as an alcohol or a builder.

Couette Flow   The flow of liquid in the annulus between two concentric cylinders that rotate at different speeds. In the Couette rheometer one cylinder rotates, and torque is measured at the other. → Rheometer.

Couette Rheometer   → Couette Flow.

Coulter Counter Technique A particle- or droplet-sizing technique in which the flow of dispersed species in a capillary, between charged electrodes, causes changes in conductivity that are interpreted in terms of the sizes of the species. Coulter is the brand name for the automated counter. → Sensing-Zone Technique.

Coulter, Wallace (Henry) An American electrical engineer best known for his invention of
(1913–1998)   the Coulter principle (in 1948) and the Coulter Counter (in the mid 1950s). The Coulter Counter is a machine used to determine the compete blood count, the most common medical diagnostic test and a foundation of haematology. The Coulter Counter technique also used to count and measure the sizes of particles and droplets in many other kinds of colloidal dispersions. → Coulter Counter Technique, → Sensing-Zone Technique. *See* Reference [83].

Counterions   In systems containing large ionic species (colloidal ions, membrane surfaces, etc.), counterions are those that, compared with the large ions, have low molecular mass and opposite charge sign. For example, clay particles are usually negatively charged and are naturally associated with exchangeable counterions such as sodium and calcium. In the early literature the term "Gegenion" was used to mean counterion. → Co-Ions.

CPC   → Condensation Particle Counter.

CPE   → Cloud Point Extraction.

Cream   The process of creaming in a dilute emulsion usually produces a discernible, more concentrated emulsion termed "cream" and having a volume termed the "cream volume".

Creaming   The process of emulsion droplets floating upwards under gravity or in a centrifugal field to form a concentrated emulsion (cream) quite distinct from the underlying dilute emulsion. This is not the same as the breaking of an emulsion. → Sedimentation.

Cream Volume   → Cream.

Creep Compliance   → Creeping Flow.

Creep Curve   The result of measuring, under a constant applied shear stress, the deformation of a material over time. → Viscoelastometer.

Creeping Flow   Gradual deformation and/or flow under an applied stress. Creep compliance is the ratio of strain to initially applied stress in a viscoelastic material. → Viscoelastometer.

Critical Aggregation   A transition concentration for hydrophobically associating poly-
Concentration   mers. Above the critical aggregation concentration the polymer solution viscosities are significantly higher than those for solutions of the equivalent non-associating polymer; below the critical aggregation concentration the polymer solution viscosities are significantly lower than those for solutions of the equivalent non-associating polymer. Also termed the "polymer overlap concentration". → Hydrophobically Associating Polymers.

Critical Coagulation   (CCC) The electrolyte concentration that marks the onset of
Concentration   coagulation of dispersed species. The CCC is very system-specific, although the variation in CCC with electrolyte composition has been empirically generalized. → Schulze-Hardy Rule. If the CCC in a binary electrolyte of salts A and B occurs at concentrations c_A and c_B compared with the pure component CCCs of c_A° and c_B°, then three types of additive effects can be distinguished [4] as follows. The electrolytes are:
additive if $(c_A/c_A^\circ) + (c_B/c_B^\circ) = 1$
antagonistic if $(c_A/c_A^\circ) + (c_B/c_B^\circ) > 1$
synergistic if $(c_A/c_A^\circ) + (c_B/c_B^\circ) < 1$

Critical Coagulation Temperature	(CCT) The minimum temperature to which a dispersion must be raised to induce coagulation. → Critical Coagulation Concentration.
Critical Deposition Velocity	The minimum flow velocity for which dispersed solids flowing in a pipe remain suspended and do not form a bed of particles along the bottom of the pipe. → Transition Velocity.
Critical Film Thickness	A fluid film may thin to a narrow range of film thicknesses within which it either becomes metastable to thickness changes (equilibrium film) or else ruptures. Persistent foams comprise fluid films at their critical film thickness.
Critical Flocculation Concentration	→ Critical Coagulation Concentration.
Critical Flocculation Temperature	→ Critical Coagulation Temperature.
Critical Micelle Concentration	(c.m.c. or CMC) The surfactant concentration above which molecular aggregates, termed "micelles", begin to form. In practice a narrow range of surfactant concentrations represents the transition from a solution in which only single, unassociated surfactant molecules (monomers) are present to a solution containing micelles. Useful tabulations are given in References [22, 23, 84, 85]. → Micelle, and *see* Table 13.
Critical Micelle Temperature	(CMT) Usually used to refer to the lowest temperature at which micelles can form in aqueous solution. The term should be used with caution since, depending on the nature of the surfactant, the critical micelle concentration (CMC) may increase or decrease with temperature, may exhibit a minimum with temperature, or may be essentially independent of temperature.
Critical Surface Tension of Wetting	The minimum, or transition, surface tension of a liquid for which it will no longer exhibit complete wetting of a solid. This value is usually taken to be characteristic of a given solid and is sometimes used as an estimate of the solid's surface tension. For a given solid it is typically determined by plotting the cosine of contact angles between the solid of interest and a series of liquids versus the surface tensions of those liquids (a Zisman plot). The surface tension extrapolated to zero contact angle is the critical surface tension of wetting of the solid. → Hydrophobic Index.
Critical Temperature	In adsorption, the transition temperature at which a monolayer no longer exhibits the properties of a condensed state.
Critical Thickness	→ Critical Film Thickness.
Critical Wetting Surface Tension	→ Critical Surface Tension of Wetting.

Cross-Flow Filtration	→ Filtration.
Cross-Sectioning	Mechanically cleaving, cutting, and/or polishing in the plane perpendicular to that being studied in order to be able to observe compositional variations in the plane.
Crude Oil	A naturally occurring hydrocarbon produced from an underground reservoir. In the petroleum field, distinctions drawn among light, heavy, extra-heavy, and bituminous crude oils are made. *See also* References [50–52], → Oil, → Light Crude Oil, → Heavy Crude Oil, → Extra-Heavy Crude Oil, → Bitumen, → Asphalt.
Cryogenic SEM	→ Freeze-Fracture Method, → Scanning Electron Microscopy.
Cryogenic TEM	→ Freeze-Fracture Method, → Transmission Electron Microscopy.
Crystalloid	→ Colloid.
CSWNT	→ Carbon Single-Wall Nanotube. → Carbon Nanotube.
CTF	Columnar Thin Film. → Sculptured Thin Film.
Cuff-Layer Emulsion	→ Interface Emulsion.
Curd-Fibers	→ Soap Curd.
Curd Soap	→ Soap Curd.
Curds	Part of the process to make cheese involves the flocculation of an electrostatically stabilized colloidal O/W emulsion of oil droplets coated with milk casein. The final aggregates of these droplets are termed curds. The remaining liquid is termed whey. Heating the curds helps to further separate-out the whey and convert the curd to an elastic solid (cheese). → Casein.
Curie Point	The temperature above which a ferromagnetic material becomes paramagnetic. → Ferromagnetic.
Cut Size	→ Elutriation.
Cutting Mill	A machine for the comminution, or size reduction, of materials. Such machines use a rotating shaft on which is mounted a series of cutting knives that interleave with a series of separately mounted stationary knives. Cutting mills can reduce materials to particles on the order of $100\,\mu m$.
CVC	Colloid Vibration Current. → Ultrasound Vibration Potential.
CVD	→ Chemical Vapour Deposition.
CVP	Colloid Vibration Potential. → Ultrasound Vibration Potential.
Cyclimetric Water	→ Polywater.

Cyclone

A separation device in which a fluid tangentially enters a conical chamber from one side, whirls around and then leaves the chamber through an opening at the top. The whirling fluid (liquid or gas) accelerates as it moves up towards the top of the chamber so that larger-sized, heavier particles or droplets eventually become unable to flow the fluid flow and hit the cyclone wall. From there the particles or droplets drop to the bottom where they are drawn off. The lighter particles or droplets follow the fluid up and through the top of the chamber. A series of cyclones can be used to separate particles or droplets into several size fractions. Elutriation and impactors operate on a similar principle.

D

Danish Agar	→ Furcellaran.
Darcy's Law	→ Permeability.
Dark-Field Illumination	A kind of illumination for microscopy in which the illumination of a specimen is arranged so that transmitted light falls out of the optical path of the microscope and only light scattered by a dispersed phase is observed. It is used to detect the presence of dispersed species that are smaller than the resolving power of the microscope. Sometimes termed dark-ground illumination. A microscope using this principle is referred to as an ultramicroscope. Example: commonly used in particle microelectrophoresis. Compare with Bright-Field Illumination.
Dark-Field Microscope	→ Ultramicroscope, → Dark-Field Illumination.
Dark-Ground Illumination	→ Dark-Field Illumination.
Dead-End Filtration	→ Filtration.
Deaeration	The removal of the gas phase from a dispersion. Example: Some nonaqueous foams (made from bitumen or heavy crude oils) are very viscous and are deaerated by processes such as contacting with steam in cascading froth, countercurrent steam-flow vessels.
Deaerator	A vessel designed to remove gas (usually air) from a dispersion.
Deborah Number	In rheology, the dimensionless ratio of relaxation time for a process to the time of observation of that process.
Debye Forces	Attractive forces between molecules due to dipole-induced dipole interaction. → van der Waals Forces.
Debye-Hückel Parameter	→ Debye Length.

Debye-Hückel Theory	A description of the behavior of electrolyte solutions in which ions are treated as point charges and their distribution is described in terms of a competition between electrical forces and thermal motion. $\rightarrow$ DLVO Theory.
Debye Length	A parameter in the Debye-Hückel theory of electrolyte solutions, κ^{-1}. For aqueous solutions at $25\,°C$, $\kappa = 3.288\sqrt{I}$ in reciprocal nanometres, where I is the ionic strength of the solution. The Debye length is also used in the DLVO theory, where it is referred to as the electric-double-layer thickness and represents the distance over which the potential falls to $1/e$, about one-third, of the value of the surface potential. Also termed the Debye Screening Length. $\rightarrow$ Electric-Double-Layer Thickness.
Debye Parameter	$\rightarrow$ Debye Length.
Debye, Peter (Joseph William) (1884–1966)	A Dutch physicist and chemist who made many contributions to colloid chemistry (light scattering and X-ray diffraction) and to physical chemistry generally (theory of specific heats, the concept of permanent molecular dipole moments). Eponyms include "Debye-Hückel theory" (conductivity of strong electrolyte solutions), "Debye-Scherer" (X-ray powder) analysis, the "Debye temperature" (in specific heats), "Debye forces" (in van der Waals forces), and the "Debye" (unit of dipole moment). Debye won the Nobel prize in chemistry (1936) for contributions to molecular structure through his investigations on dipole moments and the diffraction of X-rays and electrons in gases. *See* References [86, 87].
Deconvolution	In electron and X-ray spectroscopies, a mathematical means of eliminating the effect of a factor contributing to line width, from a peak.
Decrepitation	A phenomenon in which solid particles or liquid droplets containing voids filled with water or gas, disintegrate or explode upon heating, typically causing a crackling sound.
Deep Bed Filtration	A kind of filtration process in which a suspension flows through a bed of coarse-grained material, such as sand, and particles are removed throughout the filter bed (as opposed to only on the surface of a filter or filter cake). Also termed depth filtration, or granular media filtration. Deep bed filters operating at low approach velocities (up to about $0.5\,m/h$) are also termed slow sand filters. $\rightarrow$ Filtration.
Deflocculation	The reverse of aggregation (or flocculation or coagulation). Same as Peptization.
Defoamer	$\rightarrow$ Foam Breaker, $\rightarrow$ Antifoaming Agent.
Degree of Association	In micelles, this term refers to the number of surfactant molecules in the micelle. $\rightarrow$ Aggregation Number.

Degree of Polymerization	The number of repeating monomers in a polymer molecule.
De-Inking	De-inking is used to remove ink particles from cellulose fibres as part of the recycling of waste papers. The principal processes used are washing de-inking, flotation de-inking, or a hybrid of the two (flotation-washing de-inking). Washing de-inking involves wetting, displacing, and dispersing the ink using alkali, an oxidizing agent and a detergent. Flotation de-inking is similar but also involves the use of collectors, frothers, and pH regulators, such as are used in mineral flotation processes.
Demulsibility	The relative ease with which a particular emulsion may be broken and separated into its constituent liquid phases.
Demulsification	→ Demulsifier.
Demulsifier	(1) Chemical: Any agent added to an emulsion that causes or enhances the rate of breaking of the emulsion (separation into its constituent liquid phases). Demulsifiers can act by any of a number of different mechanisms, which usually include enhancing the rate of droplet coalescence.
	(2) Device: Any device that is used to break emulsions. Such devices may use chemical, electrical, thermal, or mechanical means, or a combination, to break an emulsion and cause separation into its constituent liquid phases.
Dendrimer	A synthetic polymer produced by adding symmetric branches to a monomer to produce a highly branched three-dimensional structure. Dendrimers are of interest in molecular nanotechnology as potential backbones to which can be attached other molecules. → Dendritic.
Dendritic	Having a highly branched, often tree-like or fern-like structure. The term has been used to apply to particles, aggregates (branched-chain aggregate), polymers (dendrimers), and mineral forms (dendrites). Example: dendrites of manganese oxide, having the appearance of ferns, are commonly found on the surfaces of limestone rock and are sometimes mistaken for fossils.
Dense Nonaqueous-Phase Liquid	(DNAPL) → Nonaqueous-Phase Liquid.
Depletion Flocculation	The flocculation of dispersed species induced by non-adsorbing polymer molecules due to depletion forces. → Depletion Force.
Depletion Force	When solutes such as non-adsorbing polymer molecules do not, for some reason, enter the gap between adjacent surfaces an attractive force is created between the surfaces. This depletion force arises out of the solute's ability to influence osmotic pressure in bulk but not in the gap between the surfaces. → Depletion Flocculation.

Depolarization	In light scattering, the reduction in polarization of light scattered at 90° from anisotropic particles as compared with that scattered from isotropic particles. → Cabannes Factor.
Depolarization Ratio	In light scattering, the ratio of intensities of light polarized horizontally to vertically.
Deposition Velocity	→ Critical Deposition Velocity.
Depressant	Any agent that can be used in froth flotation to selectively reduce the effectiveness of collectors for certain mineral components. Example: the use of cyanide to complex with copper and prevent adsorption of collectors, and therefore prevent the flotation of copper during the flotation of base metal sulphides with xanthates. → Froth Flotation.
Depth Filtration	→ Filtration.
Depth Profiling	In electron and X-ray spectroscopies, the monitoring of the signal strength of any variable that can be related to distance from the surface of a target.
Derjaguin, Boris (Vladimirovich) (1902–1994)	A Russian chemical physicist known for his research on the effects solid surfaces can exert on the properties of liquids with which they are in contact. With L.D. Landau he developed a theory of the stability of colloidal particles that could explain the Schulze-Hardy rule. Independently developed also by Verwey and Overbeek the theory became famous as the DLVO theory, an acronym constructed from the last names of these four scientists. He also worked extensively in the areas of liquid/liquid and solid/solid interactions, and introduced the concept of disjoining pressure. In some translations the last name is spelled Deryagin or Deriagin. *See* References [88, 89].
Desalter	An oil-field or refinery apparatus used to separate water and associated dissolved salts from crude oil.
Desorption	The process by which the amount of adsorbed material becomes reduced. That is, the converse of adsorption. Desorption is a different process from negative adsorption. → Adsorption.
Detection Limit	The smallest concentration of a species that can be measured under specified analytical conditions. Operationally, detection limit is frequently taken to be the concentration for which the total signal minus the background signal is two to three times the standard deviation of the background signal above the background signal.
Detergency	The action of surfactants that causes or aids in the removal of foreign material from solid surfaces by adsorbing at interfaces and reducing the energy needed to effect the removal. The processes of

removal by dissolution and removal by abrasion are not considered to be part of detergency. → Detergent.

Detergent

A surfactant that has cleaning properties in dilute solutions. As commercial cleaning products, detergents are actually formulations containing a number of chemical components, including surfactants, builders, bleaches, brighteners, enzymes, opacifiers, and fragrances. In such formulations there is usually a principal surfactant, termed the "main active surfactant", and a secondary surfactant (s), termed the "coactive surfactant(s)". The non-surface active components are termed "ancillary components", or "ancillaries".

Detergentless Microemulsion

Does not really refer to an emulsion but rather to making an otherwise insoluble component (an oil) soluble by adding a third component. See Reference [24].

Detergent Oil

A lubricating oil, formulated to contain surfactant, that has detergent properties in the sense that solid particles are dispersed and kept in suspension. Example: A detergent oil can be used in an internal combustion engine. See Reference [41].

Detersion

The action of initiating or causing detergency to happen. → Detergency, → Detergent.

Dewetting

In antifoaming, the process by which a droplet or particle of antifoaming agent enters the gas-liquid interface and displaces some of the original liquid from the interface. The liquid is usually an aqueous phase, so the process is sometimes referred to as "dewetting".

DFM

Dynamic Force Microscopy. → Scanning Probe Microscopy.

Dialysate

→ Dialysis.

Dialysis

A separation process in which a colloidal dispersion is separated from a noncolloidal solution by a semipermeable membrane, that is, a membrane permeable to all species except the colloidal-sized ones. Osmotic pressure difference across the membrane drives the separation. The solution containing the colloidal species is referred to as the "retentate" or "dialysis residue". The solution that is free of colloidal species is referred to as the "dialysate" or "permeate"; at equilibrium (no osmotic pressure difference) this solution is referred to as the equilibrium dialysate. → Filtration.

Dialysis Residue

→ Dialysis.

Diamagnetic

A material that is repelled by an external magnetic field. In contrast, paramagnetic materials are attracted into an external magnetic field, and ferromagnetic materials have magnetic properties independent of external magnetic fields.

Dielectric Constant	$\rightarrow$ Permittivity.
Dielectric Saturation	The reduction in relative permittivity due to the application of an electric field.
Dielectrophoresis	The motion of dipolar colloidal species caused by an imposed, nonhomogeneous electric field. The species move toward the most intense part of the electric field with a dielectrophoretic velocity that depends on their dipole moment, any electric charge, and the electric field intensity gradient. $\rightarrow$ Clausius-Mossotti Factor, $\rightarrow$ Levitator.
Dielectrophoretic Force	$\rightarrow$ Dielectrophoresis.
Differential Capacitance of the Electric Double Layer	$\rightarrow$ Capacitance of the Electric Double Layer.
Differential Diffusion Coefficient	$\rightarrow$ Diffusion Coefficient.
Differential Maximum Bubble Pressure Method	$\rightarrow$ Maximum Bubble Pressure Method.
Differential Mobility Analysis	(DMA) A method for the determination of size distribution in an aerosol of solid particles. DMA is based on the electrical mobility (electrophoretic mobility) of charged particles, in a carrier gas, moving in an electric field gradient. Also termed electrostatic classifier. Variations include micro-differential mobility analysis (μDMA) and nano-differential mobility analysis (nDMA).
Differential Refractometer	An instrument capable of measuring refractive index differences between materials, such as between a solution and its pure solvent. Such instruments can determine very small differences in refractive index.
Differential Viscosity	The rate of change of shear stress with respect to shear rate, taken at a specific shear rate ($\eta_D = d\tau/d\dot{\gamma}$). *See* Table 4.
Diffuse Double Layer	$\rightarrow$ Diffuse Layer, $\rightarrow$ Electric Double Layer.
Diffuse Layer	The Gouy layer, in an Electric Double Layer.
Diffusion	The spontaneous movement of species in response to a gradient in their chemical potential. Such gradients can be caused by gradients in temperature or in species concentration, for example. Fick's first and second laws specify diffusion coefficients in terms of concentration gradients and form the basis for experimental measurements of diffusion coefficients.
Diffusion Coefficient	According to Fick's first law, the diffusion coefficient (properly the differential diffusion coefficient) is the ratio of amount of species flowing through unit area in unit time, to the concentration gradient of the same species. If extrapolated to zero concentration

of diffusing species, it is the limiting diffusion coefficient. The self-diffusion coefficient is the diffusion coefficient of a species when there is no chemical potential gradient.

Diffusiophoresis

The movement of a colloidal species in response to the concentration gradient of another dissolved, noncolloidal, solute. Also termed Stefan Flow.

Diffusivity

$\rightarrow$ Diffusion Coefficient.

Dilatant

A non-Newtonian fluid for which viscosity increases as the shear rate increases. The process is termed "shear thickening". Example: Wet sand on a beach. IUPAC recommends this term be used for the sub-category of shear thickening fluids whose volume increases upon thickening. *See also* Reference [4].

Dilational Elasticity

$\rightarrow$ Film Elasticity.

Diluent

A low-boiling petroleum fraction, such as naphtha, which is added to a more viscous high- boiling petroleum liquid or oil-continuous emulsion. The diluent is usually added to reduce viscosity.

Dimple

The region of a draining liquid film that exhibits certain shapes, such as a bell-shape, rather than the interfacial curvature of the equilibrium film.

Dip Coating

A method for depositing a thin liquid film coating in which a solid substrate is pulled-out of a liquid phase, or alternatively, in which the liquid phase is allowed to flow and drain down the solid's surfaces.

Dipole

A separation of electrical charge within an electrically neutral species. This process is different from bipolar. Dipoles can be permanent or induced temporarily by an applied electric field.

Dip Pen Nanolithography

(DPN) A form of lithography, at the nanometre scale, in which molecules are delivered from a coating on an an atomic force microscope (AFM) tip, or through a capillary or nanotube at the tip of an AFM. DPN can be used to make lines of ink or glue, for example, which are tens of molecules wide and one molecule thick. With appropriate control of the tip, DPN forms the basis for a nanopen or a nanoplotter. DPN can also be used to create three-dimensional patterns, including patterns that do not etch away under selected chemical treatments, forming what are termed nanoresists. $\rightarrow$ Nanoimprint Lithography, $\rightarrow$ Nanophotolithography. *See* Reference [90].

Disassembler

(1) (Nanotechnology) A device for taking things apart through mechanical action on the nanoscale. A simple example is a device that can pick-up an atom and move it to another position, such as

can be done with an atomic force microscope. Such a device has also been termed a nanodisassembler, or a nanocrane. → Assembler. (2) On a more sophisticated level, a disassembler is a conceptual machine or material that can reduce a molecular structure to component fragments, atoms, or molecules. Also termed chemical disassembler, or molecular dissassembler. A possible application example would be the use of nanodissassemblers to treat waste by breaking down hazardous chemical molecules into less hazardous components.

Discontinuous Phase → Dispersed Phase.

Discreteness of Charge Charged colloidal species usually obtain their charge from a collection of discrete charge groups present at their surfaces. This discreteness of charge is, however, frequently approximated as a uniform surface charge distribution in descriptions of colloidal phenomena (e.g., DLVO theory). → Esin-Markov Effect.

Disjoining Pressure The negative derivative with respect to distance of the Gibbs energy of interaction per unit area yields a force per unit area between colloidal species, termed the "disjoining pressure". Example: In a thin liquid film, the disjoining pressure equals the pressure, beyond the external pressure, that has to be applied to the liquid in the film to maintain a given film thickness.

Disk Attrition Mill → Disk Mill.

Disk Mill A machine for the comminution, or size reduction, of wood products or other material. Such machines crush the input material between two grinding plates mounted on rotating disks. Also termed "disk attrition mill".

Dispersant Any species that can be used to aid in the formation of a colloidal dispersion. Examples: dispersant for dyestuffs, dispersant for pigments. Often a surfactant, such as a fatty acid derivative.

Disperse → Dispersion.

Dispersed Phase In a colloidal dispersion, the phase that is distributed, in the form of particles, droplets, or bubbles, in a second, immiscible phase that is continuous. Also referred to as the disperse, discontinuous, or internal phase. → Continuous Phase.

Disperse Phase → Dispersed Phase.

Dispersing Agent → Dispersant.

Dispersion (1) Colloids: A system in which finely divided droplets, particles, or bubbles are distributed in another phase. As it is usually used, dispersion implies a distribution without dissolution. An emulsion is an example of a colloidal dispersion; → Colloidal.

(2) Fluid-flow phenomena: The mixing of one fluid in another, immiscible fluid by convection and molecular diffusion during flow through capillary spaces or porous media.

(3) Groundwater contamination: The mixing of a contaminant with a noncontaminant phase. The mixing is due to the distribution of flow paths, tortuosity of flow paths, and molecular diffusion.

Dispersion Forces
The London interaction forces between any two bodies of finite mass. A component of the van der Waals forces. The term "dispersion" in dispersion forces comes from an analogy to the refraction (dispersion) of light due to induced dipole interactions. Since London's induced dipole-induced dipole interactions resemble this, the term "Dispersion Forces" was coined which is unfortunate in that these dispersion forces act against the dispersion of colloidal particles. → van der Waals Forces.

Dispersion Medium
The continuous phase in a dispersion.

Dispersion Methods
The class of mechanical methods used for preparing colloidal dispersions in which particles or droplets are progressively subdivided. → Condensation Methods.

Dispersion Mill
→ Colloid Mill.

Displacement Electrophoresis
→ Isotachophoresis.

Dissolved-Gas Flotation
→ Froth Flotation.

Dissolved Organic Carbon
→ Total Organic Carbon.

Disymmetry Ratio
In light scattering, the ratio of light scattered at an angle of 45° to that at 135°. The ratio is related to the size and shape of the dispersed species causing the scattering.

Division
→ Lamella Division.

DLS
Dynamic Light Scattering. → Photon Correlation Spectroscopy.

DLVO Theory
An acronym for a theory of the stability of colloidal dispersions developed independently by B. Derjaguin and L. D. Landau in one research group, and by E. J. W. Verwey and J. Th. G. Overbeek in another. The theory was developed to predict the stability against aggregation of electrostatically charged particles in a dispersion. → Gibbs Energy of Interaction, → Primary Minimum, → Short-Range Interaction Forces.

DMA
→ Differential Mobility Analysis.

DNAPL
Dense nonaqueous-phase liquid. → Nonaqueous-Phase Liquid.

DOC
→ Total Organic Carbon.

Donnan E.M.F.	→ Donnan Equilibrium.
Donnan Equilibrium	The equilibrium in a system in which one ionic solution is separated from another by a semipermeable membrane or other barrier, that is impermeable to at least one of the ionic species. The potential difference at zero current between two identical salt bridges placed into the two solutions is the Donnan potential, or Donnan E.M.F. → Membrane Potential.
Donnan Potential	→ Donnan Equilibrium.
Donnan Pressure	→ Colloid Osmotic Pressure.
Doppler Broadening	→ Doppler Effect.
Doppler Effect	The change in frequency of radiation emanating from a source that is in motion relative to the stationary position of detection. Also referred to as Doppler Broadening or Doppler Shift.
Doppler Shift	→ Doppler Effect.
Dorn Effect	→ Sedimentation Potential.
Dorn Potential	→ Sedimentation Potential.
Double Emulsion	→ Multiple Emulsion.
Double Layer	→ Electric Double Layer.
Double-Layer Thickness	→ Electric-Double-Layer Thickness.
DPN	→ Dip Pen Nanolithography.
DRA	→ Drag Reducing Agent.
Drag	The force due to friction experienced by a moving dispersed species.
Drag Reducing Agent	(DRA) Any materials that are added, in small quantity, to substantially reduce the friction factor for liquids flowing under turbulent conditions in tubes or pipes. Examples include certain polymers and surfactants. Reducing the friction factor enables increased flow rates for the same amount of pumping energy. Also termed drag reducers or flow improvers.
Draves Wetting Test	A method for comparing the wetting power of surfactants. It measures the time required for complete wetting of a piece of cloth or skein of yarn placed at the surface of a surfactant solution, under specified test conditions. Different systems are compared in terms of their wetting times. → Wetting.
Drexler, (Kim) Eric (1955–)	An American theoretical researcher; often described as the 'father of nanotechnology.' Drexler's 1981 paper "Molecular Engineering: An Approach to the Development of General Capabilities

for Molecular Manipulation" established the basic principles of molecular nanotechnology. Drexler's Ph.D. from the Massachusetts Institute of Technology was the first to be awarded in the field of molecular nanotechnology. *See* References [91, 92]. → Feynman, Richard P., → Iijima, Sumio, → Taniguchi, Norio.

Drizzle	Atmospheric Aerosols of Liquid Droplets, *see* Table 5.
Droplet	A liquid globule surrounded by another phase. The surrounding phase is usually a gas (as in a droplet of rain falling through the atmosphere) but may be another liquid (as in a droplet in an emulsion) or even a solid (as in a solid emulsion).
Dropping Mercury Electrode	An electrode comprising a capillary filled with mercury and arranged such that by forming and releasing mercury drops from the capillary tip, a renewable electrode surface can be presented to a solution under study. Used in polarography.
Drop-Volume Method	A method for determining surface or interfacial tension based on measuring the volume of drops that form at and fall from the tip of a capillary. → Drop-Weight Method.
Drop-Weight Method	A method for determining surface or interfacial tension based on measuring the mass of drops that form at and fall from the tip of a capillary. In early literature referred to as the Stalagmometric Method. → Drop-Volume Method.
Dry Foam	→ Foam.
Dry Nanotechnology	→ Wet Nanotechnology.
du Noüy, Pierre (Lecomte) (1883–1947)	A French biophysicist known for his work on the biophysical properties of serum and plasma. He is remembered to colloid and interface scientists for his introduction of the concept that the surface tension of a liquid can be determined from the maximum force required to pull a thin ring, mounted horizontally, from the surface of a liquid, now known as the *du Noüy ring method*.
du Noüy Ring Method	A method for determining surface or interfacial tension based on measuring the force needed to pull an inert ring through an interface. → Wilhelmy Plate Method.
Duplex Film	Any film that is thick enough for each of its two interfaces to be independent of each other and exhibit their own interfacial tensions. A duplex film is thus thicker than a monomolecular film.
Dust	A dust cloud is an aerosol of solid particles (dispersion of solid particles in gas) in which the particles have resulted from the mechanical disintegration of larger matter. The particle sizes are usually taken to be greater than 1 μm in diameter. → Aerosol, → Aerosol of Liquid Droplets, → Fume, *see* Table 5.

Dust Devil	A kind of dust aerosol, a dust devil is a swirling plume of rising, dusty air caused by strong heating of air near the ground surface. Dust devils tend to be short-lived.
Dynamic Foam Test	Any of several methods for assessing foam stability in which one measures the steady-state foam volume generated under given conditions of gas flow, and shearing or shaking [17, 93]. → Static Foam Test, → Foaminess.
Dynamic Force Microscopy	(DFM) → Scanning Probe Microscopy.
Dynamic Interfacial Tension	→ Equilibrium Surface Tension.
Dynamic Light Scattering	(DLS) Similar to conventional light scattering except that the coherence of a laser light beam is used to obtain additional information about the Brownian motions of dispersed species; mostly used for the size determination of spherical particles. → Photon Correlation Spectroscopy.
Dynamic Surface Tension	→ Equilibrium Surface Tension.

E

EACN	→ Equivalent Alkane Carbon Number.
EAI	→ Emulsifying Activity Index.
EC	→ Emulsifying Capacity.
Eddy Kinematic Viscosity	In turbulent flow applications the effective viscosity is increased due to increased shear stresses (internal friction) due to the increased momentum transfer caused by eddies. The eddy viscosity is a function of the flow, not of the fluid, and increases with increasing turbulence. In turbulent flow applications the kinematic viscosity is usually simply replaced by the eddy kinematic viscosity because the eddy viscosity is typically of the order of one hundred thousand times the kinematic viscosity. Also termed Kinematic Eddy Viscosity.
Eddy Viscosity	→ Eddy Kinematic Viscosity.
EDL	→ Electric Double Layer.
EDXS	→ Energy Dispersive X-Ray Spectroscopy.
EELS	Electron energy loss spectroscopy. → Characteristic Energy-Loss Spectroscopy.
Effective Particle Size	A measure or estimate of particle size based on the properties or behaviour of the particles in a particular system.
Effective Porosity	→ Porosity.
Effective Viscosity	For foams or emulsions flowing in porous media, the foam's effective viscosity is that calculated from Darcy's Law. This calculation is an approximation because foams are compressible, and foams and emulsions are usually non-Newtonian.
EIA	Energetic Ion Analysis. → Ion Beam Analysis.

Dictionary of Nanotechnology, Colloid and Interface Science. Laurier L. Schramm
Copyright © 2008 WILEY-VCH Verlag GmbH & Co. KGaA, Weinheim
ISBN: 978-3-527-32203-9

EID	→ Electron-Impact Desorption Spectroscopy.
Eigenkolloide	→ Pseudocolloid.
Eilers Equation	An empirical equation for estimating the viscosity of a dispersion. *See* Table 11.
Einstein, Albert (1879–1955)	Primarily a physicist, the German-born American Einstein is known to colloid science for contributions in the areas of Brownian movement, light scattering, sedimentation equilibrium, and viscosity. Eponyms include the "Einstein equation" (viscosity), the "Stokes-Einstein equation" (diffusion coefficient), and "Einsteinium" (element 99). Einstein is best known for his contributions to quantum theory and his theories of special relativity, and general relativity. He received the Nobel prize (1921) in physics for his work on the nature of light, particularly on the photoelectric effect. *See* References [94, 95].
Einstein Equation	(1) Diffusion: Relation between the diffusion coefficient of a dispersed species and its friction factor: $D = kT/f$, where $f = 6\pi\eta r$ (η is viscosity, r is radius). (2) Viscosity: An equation for estimating the viscosity of a dispersion. *See* Table 11.
Einstein-Smoluchowski Equation	Relation giving the root mean square displacement of a dispersed species having Brownian motion as $2Dt$, where D is diffusion coefficient and t is time.
EIS	Electron-impact spectroscopy. → Characteristic Energy-Loss Spectroscopy.
Ekman Effect	Ocean currents induced by drag from winds blowing over the surface develop downward as well as at the surface. Over time the Coriolis force changes the direction of flow, with varying change occurring with depth. The current turns further in direction, but decreases in strength, with depth. This is the Ekman Effect; it creates a spiral known as an Ekman spiral. → Coriolis Effect.
Ekman Spiral	→ Ekman Effect.
Elasticity	The ability of a material to change its physical dimensions when a force is applied to it, and then to restore its original size and shape when the force is removed. → Film Elasticity.
Elasticity Number	A dimensionless quantity (E_s) characterizing the surface-tension gradient in a thinning foam film. For systems containing only one surfactant: $E_s = -(d\gamma/d \ln \rho^s)(R_f/[\eta D])$, where γ is the surface tension, ρ^s is the surface density of the surfactant, R_f is the thin-film radius, η is the bulk liquid viscosity, and D is the diffusivity of the surfactant.

Elastic Limit	The largest force per unit area that can be applied to an elastic substance without causing irreversible deformation.
Elastic Low-Energy Electron Diffraction	(ELEED) $\rightarrow$ Low-Energy Electron Diffraction.
Elastic Scattering	$\rightarrow$ Light Scattering.
Electric Double Layer	(EDL) An idealized description of the distribution of free charges in the neighborhood of an interface. Typically the surface of a charged species is viewed as having a fixed charge of one sign (one layer), while oppositely charged ions are distributed diffusely in the adjacent liquid (the second layer). The second layer can be considered to be made up of a relatively more strongly bound Stern layer in close proximity to the surface and a relatively more diffuse layer (Gouy layer, or Gouy-Chapman layer), at greater distance.
Electric-Double-Layer Thickness	A measure of the decrease of potential with distance in the diffuse part of an electric double layer. It is the distance over which the potential falls to $1/e$, or about one-third, of the value of the surface or Stern layer potential, depending on the model used. Also termed the "Debye length".
Electric Endosmose	$\rightarrow$ Electro-osmosis.
Electroacoustical Methods	Either of the two methods for determining the electrokinetic potential of colloidal species in which either the species are detected by the sound waves generated when the species are made to move by an imposed alternating electric field (electrokinetic sonic amplitude) or the species are detected by the electric field generated when the species are made to move by an imposed ultrasonic field. $\rightarrow$ Acoustophoretic Mobility, $\rightarrow$ Electrokinetic Sonic Amplitude, $\rightarrow$ Ultrasound Vibration Potential, *see* References [45, 96].
Electroactive Solids	A kind of Smart Colloid composed of electroactive polymer material that will deflect when a current is passed through it. This provides a way to convert electrical energy into mechanical energy, and vice versa. *See also* Reference [28]. $\rightarrow$ Smart Colloids.
Electrocapillarity	Refers to the relationship between surface electric potential and surface or interfacial tension. This relationship is most evident for curved interfaces between mercury and aqueous solutions. The mercury-water interface is normally positively charged. If an electric field is applied to reduce the interfacial electric potential, then the interfacial tension typically rises to a maximum and thereafter decreases. This relationship is known as the electrocapillary curve. The part of the curve where interfacial tension increases is referred to as the "ascending branch" or "anodic branch"; the part of the

curve where interfacial tension decreases is referred to as the "descending branch" or "cathodic branch". The maximum in the curve is termed the "electrocapillary maximum" and occurs at zero net surface electric charge (the zero point of charge).

Electrocapillary Maximum	→ Electrocapillarity.
Electrocoagulation	Coagulation induced by exposing a dispersion to an alternating electric field gradient between two sacrificial metal electrodes. Electrocoagulation is apparently due to a combination of the alternating electric field and the adsorption on dispersed particles, or droplets, of ions solubilized from the electrodes. *See also* Reference [97].
Electrocratic	A dispersion stabilized principally by electrostatic repulsion.
Electrocrystallization	A kind of electrodeposition in which ions from solution become deposited on or into an electrode surface and then participate in crystallization, the building up of old crystals, or the growing of new ones at the electrode surface.
Electrodecantation	A separation process in which a colloidal dispersion is separated from a noncolloidal solution by an applied electric field together with the force of gravity. Also called Electrophoresis Convection.
Electrodeposition	The deposition of dissolved or dispersed species on an electrode under the influence of an electric field.
Electrodialysate	→ Electrodialysis.
Electrodialysis	A separation process somewhat like dialysis and ultrafiltration, in which a colloidal dispersion is separated from a noncolloidal solution by a semipermeable membrane; that is, a membrane permeable to all species except the colloidal-sized ones. Here, an applied electric field (rather than osmotic pressure or an applied pressure) across the membrane drives the separation. As in dialysis and ultrafiltration, the solution containing the colloidal species is referred to as the "retentate" or "dialysis residue". However, the solution that is free of colloidal species is referred to as "electrodialysate" rather than "dialysate" because the composition is usually different from that produced by dialysis. → Dialysis, → Filtration.
Electroendosmosis	→ Electro-osmosis.
Electro-Explosion	A method for the preparation of colloidal and/or nanoscale particles, in which a high-voltage/high-current pulse is applied to a wire in a reaction chamber. The electrical pulse causes the wire to explode, creating the particles. Also termed exploding wire aerosol generation.

Electrofiltration	Filtration combined with an electric field, usually in cake filtration and sludge dewatering. In electrofiltration electrophoretic motion drives particles toward filter grains and increases the probability of the particles being captured in the filter bed.
Electrofiltration Potential	In the electrical logging of well boreholes mud filtrate that is made to flow through a mud cake produces an electric potential gradient across the mud cake termed the electrofiltration potential. The electrofiltration potential is proportional to pressure and the electrical resistivity of the fluid, inversely proportional to the fluid viscosity, and can be used to estimate the formation pressure.
Electroformed Sieve	$\rightarrow$ Particle Size Classification.
Electrohydrodynamic Atomization	A method for the preparation of monodisperse aerosols of liquid droplets. A liquid is sprayed from a nozzle in the presence of an electric field gradient. The thin liquid jet that is produced breaks-up into monodisperse droplets, typically having diameters of a few micrometres.
Electrokinetic	A general adjective referring to the relative motions of charged species in an electric field. The motions can be either of charged, dispersed species or of the continuous phase, and the electric field can be either an externally applied field or else created by the motions of the dispersed or continuous phases. Electrokinetic measurements are usually aimed at determining Zeta Potentials.
Electrokinetic Potential	$\rightarrow$ Zeta Potential.
Electrokinetic Sonic Amplitude	(ESA) An electroacoustical method for determining the electrokinetic potential of colloidal species, which are detected by the sound waves generated when the species are made to move by an imposed alternating electric field. The acoustic pressure measured is termed the "electrokinetic sonic amplitude". This method can be applied to dispersions that do not transmit much light and are therefore unsuitable for conventional electrophoresis measurements. $\rightarrow$ Electroacoustical Methods, $\rightarrow$ Ultrasound Vibration Potential, *see* Table 16, and *see* References [45, 96].
Electron Beam Evaporator	A high vacuum physical vapour deposition technique in which electrons are accelerated from a filament towards a target rod, or a target material in a conducting crucible, causing the target material to evaporate. Used to make thin films of materials such as carbon, titanium, gold, silver, aluminum, and nickel.
Electron Energy Analyzer	In X-ray and other photoelectron spectroscopies, an instrument for the measurement of the number of electrons as a function of kinetic energy. This information is used to provide chemical

analysis: electron spectroscopy for chemical analysis (ESCA). → Photoelectron Spectroscopy.

Electron Energy Loss Spectroscopy	(EELS) → Characteristic Energy-Loss Spectroscopy.
Electron-Impact Desorption Spectroscopy	(EID) A technique for characterizing surfaces and adsorbed species in which a high- energy electron beam causes the ejection of surface ions, whose energy is measured.
Electron-Impact Spectroscopy	(EIS) → Characteristic Energy-Loss Spectroscopy.
Electron-Loss Spectroscopy	(ELS) → Characteristic Energy-Loss Spectroscopy.
Electron Microscopy	There are three principal types of electron microscopy: transmission electron microscopy (TEM), scanning electron microscopy (SEM), and scanning transmission electron microscopy (STEM). TEM is analogous to transmitted-light microscopy but uses an electron beam rather than light, and uses magnetic lenses to produce a magnified image on a fluorescent screen. In SEM a surface is scanned by a focused electron beam and the intensity of secondary electrons is measured and used to form an image on a cathode-ray tube. In STEM a surface is scanned by a very narrow electron beam that is transmitted through the sample. The intensities in the formed image are related to the atomic numbers of atoms scanned in the sample. Environmental scanning microscopy (ESEM) is a variation on SEM that employs a chamber in which samples can be kept hydrated, and which can therefore be used to study many types of colloidal dispersion [98]). Electron tomography is a technique that uses TEM to collect a series of 2-D projections of a specimen, taken at different tilt angles. The whole tilt series is then used to reconstruct a 3-D image of the specimen. → Field Emission Microscopy, → Field Ion Microscopy, → Scanning Probe Microscopy.
Electron Paramagnetic Resonance Spectroscopy	(EPR) Measurement of the frequency of applied energy needed to induce resonance in the energy levels occupied by unpaired electrons in a magnetic field. The resonance frequency depends on the local environment of the electrons, hence on molecular structure. Also referred to as "electron spin resonance" or "ESR spectroscopy".
Electron Probe Microanalysis	(EPMA) Surface-chemical elemental analysis based on electron-excited X-ray spectrometry. The spectrometric technique can be either wavelength dispersive X-ray spectrometry (WDXS) or energy dispersive X-ray spectrometry (EDXS). The use of a focused electron probe permits spatial resolution to sub-micrometre dimensions. → Energy Dispersive X-Ray Spectroscopy, *see* Table 7.

Electron Pump	(Nanotechnology) An electron pump comprises two quantum dots connected to separate leads and separated by a thin layer of junction material (patch) that normally acts as an insulator but which allows an electron to pass through only under certain conditions, such as when an electric pulse is applied. An electron pump permits very precise current control and/or electron counting by pumping, or allowing to pass through, electrons one-by-one.
Electron Spectroscopy for Chemical Analysis	(ESCA) → Photoelectron Spectroscopy.
Electron Spin Resonance Spectroscopy	(ESR) → Electron Paramagnetic Resonance Spectroscopy.
Electron-Stimulated Desorption Spectroscopy	(ESD) → Photon-Stimulated Desorption Spectroscopy.
Electron Tomography	→ Electron Microscopy.
Electro-Osmosis	The motion of liquid through a porous medium caused by an imposed electric field. The term replaces the older terms "electrosmosis", "electric endosmose", and "electroendosmosis". The liquid moves with an electro-osmotic velocity that depends on the electric surface potential in the stationary solid and on the electric field gradient. The electro-osmotic volume flow is what runs through the porous plug and is usually expressed per unit electric field strength. The electro-osmotic pressure is the difference across the porous plug that is required to just stop electro-osmotic flow. This method can be used to determine zeta potential. *See also* Table 16.
Electro-Osmotic Pressure	→ Electro-Osmosis.
Electro-Osmotic Velocity	→ Electro-Osmosis.
Electro-Osmotic Volume Flow	→ Electro-Osmosis.
Electrophoresis	The motion of colloidal species caused by an imposed electric field. The term replaces the older term "cataphoresis". The species move with an electrophoretic velocity that depends on their electric charge and the electric field gradient. The electrophoretic mobility is the electrophoretic velocity per unit electric field gradient and is used to characterize specific systems. An older synonym, no longer in use, is kataphoresis. The term "microelectrophoresis" is sometimes used to indicate electrophoretic motion of a collection of particles on a small scale. Previously, microelectrophoresis was used to describe the measurement techniques in which electrophoretic mobilities (hence zeta potentials) are determined by observation through a microscope. The recommended term for these latter techniques is now microscopic electrophoresis. *See* Reference [4] and *see* Table 16.

Electrophoresis Convection	→ Electrodecantation.
Electrophoretic Mobility	→ Electrophoresis.
Electrophoretic Relaxation	When a charged particle or droplet undergoes electrophoresis, it tends to move somewhat ahead of its counterions and associated liquid. This movement causes a distortion in the symmetry of the electric double layer and also introduces a local electric field gradient acting in opposition to the external field. This local electric field gradient produces a retarding force on the particle or droplet: the electrophoretic relaxation. → Electrophoretic Retardation.
Electrophoretic Retardation	The effect of an electric field gradient on counterions in the electric double layer surrounding a charged species. When a charged particle or droplet undergoes electrophoresis, its counterions and associated liquid are simultaneously driven to move in the opposite directions. Thus, a retarding force is produced on the particle or droplet: the electrophoretic retardation. → Electrophoretic Relaxation.
Electrophoretic Velocity	→ Electrophoresis.
Electro-Rheological Colloids	(ER Colloids) Dispersions whose viscosity can change (by orders of magnitude) when exposed to a sufficiently large electric field. The electric field creates induced dipoles in the particles, causing the particles to form chains and structures. This in turn causes a fluid suspension to become solid. The transition is reversible; the viscosity returns to low levels when the external field is removed. The electro-rheological effect is also termed the Winslow effect. Example: certain suspensions of charged polymer particles in silicone oil. Example: cornstarch dispersed in corn oil and subjected to an electric field gradient of about 10 000 V/cm. → Smart Colloids, *see* Reference [99].
Electrosmosis	→ Electro-Osmosis.
Electrospinning	A nanostructure fabrication process in which an electric field gradient is used to control the deposition of nanofibres onto a substrate surface.
Electrostatic Classifier	→ Differential Mobility Analysis.
Electrostatic Spray-Assisted Vapour Deposition	(ESAVD) A coating method in which pre-atomized droplets are sprayed through an electric field gradient in which the droplets undergo chemical reaction enroute to the surface.
Electrostatic Treater	A vessel used to break emulsions by promoting coalescence through the application of an electric field. → Treater.

Electrosteric Stabilization	The stabilization of a dispersed species by a combination of electrostatic and steric repulsions. An example is the stabilization of suspended solids by adsorbed polyelectrolyte molecules.
Electroviscous Effect	Any influence of electric double layer(s) on the flow properties of a fluid. The primary electroviscous effect refers to an increase in apparent viscosity when a dispersion of charged colloidal species is sheared. The secondary electroviscous effect refers to the increase in viscosity of a dispersion of charged colloidal species, which is caused by their mutual electrostatic repulsion (overlapping of electric double layers). An example of the tertiary electroviscous effect would be for polyelectrolytes in solution where changes in polyelectrolyte molecule conformations and their associated effect on solution apparent viscosity occur.
ELEED	Elastic low-energy electron diffraction. $\rightarrow$ Low-Energy Electron Diffraction.
Ellipsometry	The subject concerned with the behavior of light when it passes through or is reflected by an interface. In ellipsometry an incident beam of plane-polarized light is caused to be reflected from a coated surface and the degree of elliptical polarization produced is measured. This technique allows the thickness of the surface coating to be determined.
Elliptical Jet Method	$\rightarrow$ Oscillating Jet Method.
Elongation Shape Factor	The ratio of the length to the width of a rectangle having two sides parallel to the longest dimension of a particle. $\rightarrow$ Aspect Ratio.
ELS	Electron-loss spectroscopy. $\rightarrow$ Characteristic Energy-Loss Spectroscopy.
Eluant	$\rightarrow$ Elution.
Eluate	$\rightarrow$ Elution.
Elution	Removal of an adsorbed species from a porous medium or chromatographic column by the action of a stream of flowing gas or liquid (the eluant). The resulting solution is termed eluate.
Elutriation	The separation of smaller sized, lighter particles from larger sized, heavier particles due to the upward flow of surrounding fluid that tends to "carry" the lighter particles. For a given set of operating conditions, particles smaller than a certain size, called the cut size, will be lifted up and away. Particles larger than the cut size will sediment out. Impactors and cyclones operate on a similar principle.
Embryo	In colloid science, an aggregate of a small number of species. A critical embryo has a size corresponding to maximum Gibbs

energy (constant temperature and pressure). A larger embryo is referred to as a "homogeneous nucleus". *See* Reference [4].

Emmett, Paul Hugh (1900–1985)	An American surface chemist known for his work in heterogeneous catalysis, adsorption, and nuclear chemistry. He is particularly known for his work on a method for measuring the surface areas of porous catalysts and finely divided solids. This work involves the BET (Brunauer-Emmett-Teller) theory of multilayer adsorption. *See* References [100, 101].
Emollient	An agent that lends a soft texture to skin, hair, or membrane tissues. Used in formulated personal care products such as skin creams and hair conditioners.
Emulsator	A device designed to permit observation of the conditions under which emulsion inversion occurs.
Emulsibility	The relative ease with which two immiscible liquids can be made into an emulsion.
Emulsifiable Concentrates	Emulsion concentrates in which active ingredients are dissolved in a solvent that is in turn emulsified into an aqueous phase. When water is added to them they remain emulsified (O/W) or, in some cases, spontaneously form a microemulsion [102]. Example: some agrochemicals, such as pesticides, are formulated as emulsifiable concentrates. → Suspension Concentrates.
Emulsifier	Any agent that acts to stabilize an emulsion. The emulsifier can make it easier to form an emulsion and to provide stability against aggregation and possibly coalescence. Emulsifiers are frequently but not necessarily surfactants.
Emulsify	→ Emulsifier.
Emulsifying Activity Index	(EAI) A means of estimating the effectiveness of a surfactant in stabilizing emulsions. This test involves making an emulsion, determining the average droplet size, assuming that all of the surfactant is adsorbed in a monolayer at the interface, and calculating the area covered per amount of surfactant. *See also* Reference [103] and → Emulsifying Capacity.
Emulsifying Capacity	(EC) A means of estimating the effectiveness of a surfactant in stabilizing emulsions. This test involves dissolving a known amount of surfactant in water, adding oil, and mixing in a blender to form a primary emulsion. Oil is then added until the emulsion inverts, which provides an estimate of the amount of oil that can be emulsified by the specific amount of surfactant used. This test requires standardization of the various test conditions. *See also* Reference [78] and → Emulsifying Activity Index.

Emulsion	A dispersion of droplets of one liquid in another, immiscible liquid, in which the droplets are of colloidal or near-colloidal sizes. The term can also refer to colloidal dispersions of liquid crystals in a liquid. Emulsions were previously referred to as emulsoids, meaning emulsion colloids. $\rightarrow$ Macroemulsion, $\rightarrow$ Miniemulsion, $\rightarrow$ Microemulsion, $\rightarrow$ Nanoemulsion.
Emulsion Film	$\rightarrow$ Liquid Film.
Emulsion Flotation	A variation on standard froth flotation in which small-sized particles become attached to the surfaces of oil droplets (carrier droplets). The carrier droplets attach to the air bubbles and the combined aggregates of small desired particles, carrier droplets, and air bubbles float to form the froth. Example: emulsion flotation of very fine diamond particles with isooctane. $\rightarrow$ Carrier Flotation, $\rightarrow$ Floc Flotation, $\rightarrow$ Oil-Assisted Flotation, $\rightarrow$ Roughing Flotation, $\rightarrow$ Scalping Flotation, $\rightarrow$ Scavenging Flotation, $\rightarrow$ Froth Flotation.
Emulsion Food Products	$\rightarrow$ Margarine, $\rightarrow$ Spreadable Fats.
Emulsions of Emulsions	$\rightarrow$ Multiple Emulsion.
Emulsion Polymerization	A polymerization reaction that takes place in one phase of an emulsion. Originally used in the synthesis of latex. Example: emulsification of a monomer in aqueous solution and polymerization being initiated by adding a water-soluble initiator. $\rightarrow$ Polymer Colloid, $\rightarrow$ Microemulsion Polymerization.
Emulsion Test	In general, emulsion tests range from simple identifications of emulsion presence and volume to detailed component analyses. The term frequently refers simply to the determination of sediments in an emulsion or oil sample. $\rightarrow$ Basic Sediment and Water.
Emulsion Treater	$\rightarrow$ Treater.
Emulsoid	An older term meaning "emulsion colloid". $\rightarrow$ Emulsion.
Encapsulation	$\rightarrow$ Microencapsulation.
Endcapped Polymers	$\rightarrow$ Hydrophobically Associating Polymers.
Energetic Ion Analysis	(EIA) $\rightarrow$ Ion Beam Analysis.
Energy Dispersive X-Ray Spectroscopy	(EDXS) Bombarding an atom in a sample with an electron beam causes energy transfer to an inner orbital electron. The orbital electron is ejected from the atom and an X-ray photon is generated when the vacancy is filled from within the atom. This happens in all instruments in which samples are bombarded by electrons (including TEM and SEM). The X-ray photons generated are characteristic of the element from which they are emitted. The

intensity of the X-ray photons is proportional to the number of atoms generating them. EDXS can therefore be used in qualitative and quantitative surface analysis, often in conjunction with electron microscopy. If wavelength, rather than enrgy is measured, the technique is termed wavelength dispersive X-ray spectroscopy (WDXS). $\rightarrow$ Photoelectron Spectroscopy.

Energy of Adhesion	$\rightarrow$ Work of Adhesion.
Energy of Cohesion	$\rightarrow$ Work of Cohesion.
Energy of Immersional Wetting	$\rightarrow$ Work of Immersional Wetting.
Energy of Separation	$\rightarrow$ Work of Separation.
Energy of Spreading	$\rightarrow$ Work of Spreading.

Engineered Nanoparticles   Nanoparticles that have been designed and manufactured to have specific properties and/or specific composition.

Engler Viscosity   A parameter intended to approximate fluid viscosity and measured by a specific kind of orifice viscometer. The measurement unit is Engler seconds.

Engulfment   The process in which a particle dispersed in one phase is overtaken by an advancing interface and surrounded by a second phase. Example: when a freezing front (the interface between a solid and its freezing liquid phase) overtakes a particle, the particle will either be pushed along by the front or else be engulfed by the front, depending on its interfacial tensions with the solid and with the liquid. $\rightarrow$ Freezing Front Method.

Enhanced Oil Recovery   The third phase of crude oil production, in which chemical, miscible or immiscible gas, or thermal methods are applied to restore production from a depleted reservoir. Also known as "tertiary oil recovery". $\rightarrow$ Primary Oil Recovery, $\rightarrow$ Secondary Oil Recovery.

Enmeshment   Sweep Flocculation.

Entering Coefficient   A measure of the tendency for an insoluble agent to penetrate, or "enter", an interface (usually gas-liquid or liquid-liquid). It is -1 times the Gibbs free energy change for this process, so that entering is thermodynamically favored if the entering coefficient is greater than zero. In a gas-liquid system containing such an agent A, a liquid L, and gas, the entering coefficient is given by $E = \gamma_L^\circ + \gamma_{L/A} - \gamma_A^\circ$ where γ_L° and γ_A° are surface tensions and $\gamma_{L/A}$ is interfacial tension. When equilibria at the interfaces are not achieved instantaneously, reference is frequently made to the initial entering coefficient and final (equilibrium) entering coefficient. $\rightarrow$ Spreading Coefficient.

Enthalpy Stabilization	$\rightarrow$ Steric Stabilization.
Entropic Stabilization	$\rightarrow$ Steric Stabilization.
Envelope Volume	The volume of a particle that would be obtained by tightly shrinking a film to contain it.
Environmental Scanning Microscopy	(ESEM) $\rightarrow$ Electron Microscopy.
Eötvös Equation	A relation for predicting the variation of surface tension with temperature: $\gamma(M/\rho)^{2/3} = k(T_c - T)$, where M is the molecular mass, ρ is the density, T_c is the critical temperature of the liquid, and k is termed the Eötvös constant. *See* Table 8.
Eötvös Number	$\rightarrow$ Bond Number.
Epiphaniometer	An instrument used to measure the Fuchs surface area (active surface area) of aerosol particles. An aerosol is passed through a charging chamber where lead isotopes created from a decaying actinium source become attached to the particle surfaces. The particles are collected on a filter, from which the measured radioactivity is taken to be proportional to the Fuchs surface area.
Epitaxy	The growth of crystalline material on the surface of a different material where the substrate orients the new crystal growth.
EPMA	Electron Probe Microanalysis.
EPR	$\rightarrow$ Electron Paramagnetic Resonance Spectroscopy.
Equation of Capillarity	$\rightarrow$ Young-Laplace Equation.
Equilibrium Contact Angle	The contact angle that is measured when all contacting phases are in equilibrium. The term arises because either or both of the advancing or receding contact angles can differ from the equilibrium value. It is essential to state which interfaces are used to define the contact angle. $\rightarrow$ Contact Angle.
Equilibrium Dialysate	$\rightarrow$ Dialysis.
Equilibrium Film	$\rightarrow$ Fluid Film.
Equilibrium Interfacial Tension	$\rightarrow$ Equilibrium Surface Tension.
Equilibrium Spreading Coefficient	$\rightarrow$ Spreading Coefficient.
Equilibrium Surface Tension	Surface or interfacial tensions can change dynamically as a function of the age of the surface or interface. Thus the dynamic (pre-equilibrium) tensions are distinguished from the limiting, or equilibrium, tensions.

Equivalent Alkane Carbon Number	(EACN) Each surfactant, or surfactant mixture, in a reference series will produce a minimum interfacial tension (IFT) when measured against a different n-alkane. For any crude oil or oil component, a minimum IFT will be observed against one of the reference surfactants. The EACN for the crude oil refers to the n-alkane that would yield minimum IFT against that reference surfactant. The EACN thus allows predictions to be made about the interfacial tension behavior of a crude oil in the presence of surfactant. *See* References [104, 105].
Equivalent Diameter	$\rightarrow$ Equivalent Spherical Diameter.
Equivalent Film Thickness	Refers to an experimentally determined fluid film thickness; the term "equivalent" refers to certain assumptions about the structure and properties of the film that have been made. The experimental technique used should also be stated when using this term.
Equivalent Spherical Diameter	The equivalent diameter of an imaginary spherical particle, droplet, or bubble that behaves the same with regard to some physical property as the species under examination. Some of the physical properties used include aerodynamic, diffusion, hydrodynamic, mobility, and dimension (perimeter, surface area, and volume). Also termed equivalent diameter. $\rightarrow$ Aerodynamic Diameter; $\rightarrow$ Stokes Diameter, $\rightarrow$ Z-average Mean.
ER Colloid	Also termed ER Fluid. $\rightarrow$ Electro-Rheological Colloid.
Ertl, Gerhard (1936–)	A German surface chemist best known for his work in gas adsorption on well-defined surfaces and the relationships among the related surface structures and energetics to industrial processes. He was awarded the 2007 Nobel prize in Chemistry for "groundbreaking studies in surface chemistry," including his studies of fundamental molecular processes at the gas-solid interface.
ESA	$\rightarrow$ Electrokinetic Sonic Amplitude.
ESAVD	Electrostatic Spray-Assisted Vapour Deposition.
ESCA	Electron spectroscopy for chemical analysis. $\rightarrow$ Photoelectron Spectroscopy.
ESD	Electron-stimulated desorption spectroscopy. $\rightarrow$ Photon-Stimulated Desorption Spectroscopy.
ESEM	Environmental Scanning Electron Microscopy. $\rightarrow$ Electron Microscopy.
Esin-Markov Effect	The change in zero point of charge of a species that occurs when the electrolyte can become specifically adsorbed. In the presence of indifferent electrolytes, the zero point of charge is a constant.

ESR Spectroscopy	→ Electron Spin Resonance Spectroscopy.
Evanescent Foam	A transient foam that has no thin-film persistence and is therefore very unstable. Such foams exist only where new bubbles can be created faster than existing bubbles rupture. Examples: air bubbles blown rapidly into pure water; the foam created when a champagne bottle is opened.
EXAFS	→ Extended X-Ray Absorption Fine Structure Spectroscopy.
Excess Quantities	→ Gibbs Surface.
Excipient	Any of the inactive constituents of a formulated pharmaceutical product. Their functions may include any of "bulking-up" the product to make it easier to handle, binding together other ingredients, thickening or gelling, lubricating, and coating to improve shelf-life and/or to improve taste. Examples: lactose, magnesium stearate, carboxymethylcellulose.
Excluded Volume	The volume in a system, or near an interface, that is not accessible to molecules or dispersed species because of the presence of other species in that volume. → Free Volume.
Expansion Factor	In foaming, the ratio of foam volume produced to the volume of liquid used to make the foam. Also termed the "expansion ratio".
Expansion Ratio	→ Expansion Factor.
Exploding Wire Aerosol Generation	→ Electro-Explosion.
Ex Situ	In science and engineering, *ex situ* generally refers to an aspect of a reaction or process taking place away from where it normally occurs or was created. → *In Situ*.
Extended Surfactant	Surfactants that include groups of intermediate polarity placed between the hydrophobic tail and the hydrophilic head groups. This causes the surfactant molecule to extend further than it would otherwise into both oil and aqueous phases. Examples of intermediate polarity groups in such surfactants include ethoxy and/or polypropylene oxide chains. Extended surfactants have been used, for example, to solubilize very hydrophobic oils.
Extended X-Ray Absorption Fine Structure Spectroscopy	(EXAFS) A technique for studying the separation distance of surface atoms; it is related to X-ray photoelectron spectroscopy and is based on the effect of backscattering. X-ray absorption is determined as a function of the energy of the incident X-ray beam. A related technique is surface-extended X-ray absorption fine structure spectroscopy (SEXAFS). *See also* Table 7.
Extender Flotation	→ Oil-Assisted Flotation.

External Phase $\rightarrow$ Continuous Phase.

External Surface   When a porous medium can be described as consisting of discrete particles, the outer surface of the particles is termed the "external surface". $\rightarrow$ Internal Surface.

Extinction Coefficient $\rightarrow$ Absorbance.

Extra-Heavy Crude Oil   A naturally occurring hydrocarbon with a viscosity less than 10,000 mPa·s at ambient deposit temperature, and a density greater than 1000 kg/m^3 at 15.6 °C. *See* References [50–52].

F

Fann Viscometer	A commercial instrument brand-name that has become a general use term; better known than the Saybolt type. The basic model is a direct torque-reading, mechanically simple viscometer of the concentric-cylinder type, that is used under laboratory and field conditions for determining the viscosities of materials over a (usually limited) range of shear rates. Similar to the Brookfield Viscometer (another commercial brand).
Faraday, Michael (1791–1867)	A British physicist and chemist, Faraday is best known for contributions in electromagnetism, but also for experimental studies in many other areas including thin metal films, aerosols, hydrosols, and gels, including the light scattering properties of gold hydrosols and what is now known as the "Tyndall effect". The unit Faraday equals the total charge of Avogadro's number (one mole) of electrons. *See* Reference [106].
Faraday Sol	A very fine sol of gold particles (whose diameters can be on the order of nanometres) prepared by the reduction of a gold chloride solution with phosphorus. A procedure is given in Reference [107].
Fatty Acid Soaps	A class of surfactants comprising the salts of aliphatic carboxylic acids having hydrocarbon chains of between 6 and 20 carbon atoms. Fatty acid soaps are no longer restricted to molecules with origins in natural fats and oils. *See* Table 14.
Fatty Alcohol Surfactants	The class of primary alcohol surfactants having hydrocarbon chains of between 6 and 20 carbon atoms. Fatty alcohol surfactants are no longer restricted to molecules having their origins in natural fats and oils. *See* Table 14.
FBRM	→ Focused Beam Reflectance Measurement.

Dictionary of Nanotechnology, Colloid and Interface Science. Laurier L. Schramm
Copyright © 2008 WILEY-VCH Verlag GmbH & Co. KGaA, Weinheim
ISBN: 978-3-527-32203-9

FDS	Flash desorption spectroscopy. $\rightarrow$ Temperature-Programmed Reaction Spectroscopy.
FEM	$\rightarrow$ Field Emission Microscopy.
Feret's Diameter	A statistical particle diameter; the length of a line drawn parallel to a chosen direction and taken between parallel planes drawn at the extremities on either side of the particle. This diameter is thus the maximum projection of the particle onto any plane parallel to the chosen direction. The value obtained depends on the particle orientation, thus these measurements have significance only when a large enough number of measurements are averaged together. Example: Feret's diameter is used to calculate a particle's aspect ratio. $\rightarrow$ Martin's Diameter.
Ferrofluid	A dispersion of finely divided magnetic particles in a liquid, stabilized by electrostatic and/or steric repulsion. Example: Fe_3O_4 particles in water. $\rightarrow$ Magneto-Rheological Colloids, $\rightarrow$ Smart Colloids.
Ferrofluid Foam	Liquid foams containing a stable suspension of magnetic particles, and whose stability and structure can be modified by the application of an external magnetic field [108]. $\rightarrow$ Ferrofluid.
Ferrography	A method used to determine machine health by quantifying and examining ferrous wear particles suspended in samples of lubricating or hydraulic fluid. $\rightarrow$ Analytical Ferrography and *see* Reference [37].
Ferromagnetic	A material that achieves a high degree of magnetization in a weak magnetic field and which increases with increasing magnetic field strength. Ferromagnetic materials also exhibit residual magnetism in the absence of an external magnetic field. Above the Curie point ferromagnetic materials behave like paramagnetic materials. $\rightarrow$ Paramagnetic.
Feynman, Richard P. (1918–1988)	An American physicist who won the Nobel prize in 1965 for his role in the development of the theory of quantum electrodynamics. To Feynman is attributed the concept of what is now called nanotechnology, as expressed in his famous lecture "There's Plenty of Room at the Bottom" at the annual meeting of the American Physical Society, Dec. 1959 (published in 1960 [109]). The coining of the term nanotechnology itself is attributed to Taniguchi. Feynman is also known for his theory of superfluidity in liquid helium and his prediction that protons and neutrons are not elementary particles (now known to be composed of quarks). $\rightarrow$ Drexler, (Kim) Eric $\rightarrow$ Iijima, Sumio $\rightarrow$ Taniguchi, Norio.
FFF	$\rightarrow$ Field-Flow Fractionation.

FFFP	$\rightarrow$ Film Forming Fluoroprotein Foam.
FFM	$\rightarrow$ Friction Force Microscopy; $\rightarrow$ Scanning Probe Microscopy.
FHH Isotherm	$\rightarrow$ Frenkel-Halsey-Hill Isotherm.
Fibrillar Colloid	$\rightarrow$ Colloidal.
Fick's First Law	$\rightarrow$ Diffusion Coefficient.
Field Emission Microscopy	(FEM) A type of electron microscopy in which electrons are emitted from the charged hemispherical tip of a metal wire. The electron beam is detected at a hemispherical fluorescent screen and used to form a highly magnified image that can be used to elucidate the crystal structure of the metal tip. $\rightarrow$ Electron Microscopy, $\rightarrow$ Field Ion Microscopy, *see* Table 7.
Field-Flow Fractionation	(FFF) A method of separating particles from a dispersion flowing in a channel by applying a field perpendicular to the flow. The applied field can be a transverse flow or a thermal gradient.
Field Ion Microscopy	A variation of field emission microscopy in which gas molecules in the vicinity of a positively charged, fine metal tip lose an electron. The resulting positive ions accelerate away to strike a fluorescent screen where they are detected and used to form an image of the crystal structure of the metal tip. In this technique individual atoms can be resolved. $\rightarrow$ Electron Microscopy, $\rightarrow$ Field Ion Microscopy, *see* Table 7.
Filler	Fine-grained, inert material that is added to paper, paint, rubber, resin, etc., to improve their properties in some way.
Film	Any layer of material that covers a surface and is thin enough to not be significantly influenced by gravitational forces. $\rightarrow$ Monolayer Adsorption, $\rightarrow$ Duplex Film, $\rightarrow$ Liquid Film. A film bounded by two identical phases is termed a "symmetric film", whereas if it is bounded by two different phases it is an "asymmetric film". Several distinctions are made based on the lateral dimensions of a film: Nanoscopic, or nanometre thick film (0.1 to 100 nm) Microscopic, or micrometre thick film (0.1 to 100 μm) Macroscopic, or millimetre thick film (0.1 to 100 mm)
Film Balance	A shallow trough that is filled with a liquid, and on top of which is placed material that can form a monolayer. The surface area available can be adjusted by moveable barriers, and, by means of a float, any surface pressure thus created can be measured. Also called "Langmuir film balance", "Langmuir trough", "hydrophil balance", and "Pockels-Langmuir-Adam-Wilson-McBain trough", or "PLAWM trough".

Film Compressibility	The ratio of relative area change to differential change in surface tension. → Film Elasticity.
Film Drainage	The drainage of liquid from a lamella of liquid separating droplets or bubbles of another phase (i.e., in a foam or emulsion). Also termed "thin-film drainage". → Fluid Film.
Film Elasticity	The differential change in surface tension with relative change in area. Also termed "surface elasticity", "dilational elasticity", "areal elasticity", "compressional modulus", "surface dilational modulus", or "modulus of surface elasticity". In these terms "dilational" is sometimes written "dilatational". For fluid films the surface tension of one surface is used. The Gibbs film (surface) elasticity is the equilibrium value. If the surface tension is dynamic (time-dependent) in character then, for nonequilibrium values, the term "Marangoni film (surface) elasticity" is used. The compressibility of a film is the inverse of the film elasticity.
Film Element	Any small, homogeneous region of a thin film. The film element includes the interfaces.
Film Flotation	(1) Separation. In film flotation a dry mixture of particles is placed onto an aqueous surface. Hydrophilic particles become wetted and sediment down. The hydrophobic particles, which continue to float on the aqueous surface, can be skimmed-off and collected. → Hydrophobic Index.
	(2) Flotation. A sink/float test in which a thin layer of particles are placed on the surfaces of a series of tubes containing liquids of varying surface tension. By finding the transition point, beyond which floating particles sink, or vice versa, the critical surface tension of wetting for the particles is obtained.
Film Forming Fluoroprotein Foam	(FFFP) A fire-extinguishing foam based on very low-surface-tension producing fluouroprotein surfactants. Used in sprinkler systems and as a rapidly spreading foam on hydrocarbon fires. → Fluoroprotein Foam, → Aqueous Film Forming Foam, → Alcohol Resisting Aqueous Film Forming Foam, → Fire Extinguishing Foam.
Film Line Tension	→ Line Tension.
Film Pressure	The pressure, in two dimensions, exerted by an adsorbed monolayer. It is formally equal to the difference between the surface tension of pure solvent and that of the solution of adsorbing solute. It can be measured by using the film balance. → Film Balance.
Film Tension	An expression of surface tension applied to thin liquid films that have two equivalent surfaces. The film tension is twice the surface tension.

Film Water	In soil science, the film of water that remains, surrounding soil particles, after drainage. This layer can range from several to hundreds of molecules thick and comprises water of hydration plus water trapped by capillary forces.
Filter Cake	$\rightarrow$ Filtration.
Filter Efficiency	A measure of a filter's ability to trap and retain particles of a specified size.
Filter Ripening	In water filtration, the process in which deposition of an initial layer of particles causes the filter surface to take on a nature more similar to the particles to be removed. This process enhances the filtering (hence, removal) of the particles.
Filtration	The process of removing particles or large molecules from a fluid phase by passing the fluid through some medium that will not permit passage of the particles or large molecules. Filtration frequently refers to the removal of solid particles from suspensions. The two main types of filtration process are deep bed filtration and membrane filtration. Deep bed filtration involves passing a suspension through a bed of granular material such as fine sand. This is also known as depth filtration, granular media filtration, or slow sand filtration. Membrane filtration involves passing a suspension through a thin, porous membrane, which is usually polymer or ceramic in nature, but could also be woven fabric or metal fibres. Membrane filtration processes are often categorized by the pore sizes of the filters: microfiltration (100–3000 nm pores); ultrafiltration (10–200 nm pores); nanofiltration (0.5–10 nm pores); and hyperfiltration (<2 nm pores). Hyperfiltration is also termed reverse osmosis. Membrane filters can be operated in one of two modes: dead-end filtration, in which all of the feed passes through the filter, and cross-flow filtration, in which the feed flows across the membrane filter and only some of it passes through. Either mode can be subject to a build-up of filtered particles on the membrane surface, called a filter cake.
FIM	$\rightarrow$ Field Ion Microscopy.
Final Spreading Coefficient	$\rightarrow$ Spreading Coefficient.
Fine Droplet	A droplet having one or more dimensions in a specified size range. Fine droplets are usually defined as ranging from 100 nm to 2.5 μm, or from 100 nm to 10 μm. Sometimes the same subcategories as have been defined for fine particles are also used for droplets. Droplets in the size range 1 to 100 nm are termed nanodroplets, or ultrafine droplets. $\rightarrow$ Fine Particle.

Fine Particle	A particle having one or more dimensions in a specified size range. Fine particles are usually defined as ranging from 100 nm to 2.5 µm, or from 100 nm to 10 µm. Some common subcategories of fine particles (particulate matter, PM) are designated as follows: PM_1 refers to particles that are <1 µm, $PM_{2.5}$ for particles <2.5 µm, and PM_{10} for particles <10 µm. Depending on the classification system used, particles < 100 nm in size may be included or excluded from the above definitions. Particles in the size range 1 to 100 nm are termed nanoparticles, or ultrafine particles.
Fine Sand	$\rightarrow$ Sand, *see* Table 12.
Fire Extinguishing Foam	Foam that has been specially formulated to spread over a fire's fuel and then kill a fire by separating flames from the fuel's surface, blanketing the burning fuel to smother the fire, and cooling the fuel. Typically made for use on hydrocarbon fuel fires, these foams are often formulated with fluorocarbon surfactants, which are less soluble than hydrocarbon surfactants in hydrocarbon fuels. Examples include Fluoroprotein Foam, Film Forming Fluoro-protein Foam, Aqueous Film Forming Foam, Alcohol Resisting Aqueous Film Forming Foam, Protein Foam.
Flame Pyrolysis	A fine particle preparation method in which flame heat is used to vaporize the feedstock and initiate the gas-phase chemical reactions that produce the particles. This method can be used to prepare colloidal and/or nanoparticles.
Flash Desorption Spectroscopy	(FDS) $\rightarrow$ Temperature-Programmed Reaction Spectroscopy.
Flexibilizer	$\rightarrow$ Plasticizer.
Floc	A loose cluster of particles. $\rightarrow$ Aggregation, $\rightarrow$ Flocculation.
Flocc	$\rightarrow$ Floc.
Floc Flotation	A variation on standard froth flotation in which small sized particles aggregate into flocs that in turn attach to air bubbles. Polymers are often used to induce the flocculation. Example: the floc-flotation of coal fines. $\rightarrow$ Carrier Flotation, $\rightarrow$ Emulsion Flotation, $\rightarrow$ Roughing Flotation, $\rightarrow$ Scalping Flotation, $\rightarrow$ Scavenging Flotation, $\rightarrow$ Froth Flotation.
Flocculation	$\rightarrow$ Aggregation. The products of the flocculation process are referred to as flocs (usually), or as floccules or floccs. $\rightarrow$ Selective Flocculation.
Flocculation Value	$\rightarrow$ Critical Coagulation Concentration.
Floccule	$\rightarrow$ Flocculation.

Floc Point	The temperature below which wax or solids separate from an oil. Example: the separation of wax from kerosene used in lamps or refrigerant used in air conditioners.
Flory, Paul (John) (1910–1985)	An American industrial and academic physical chemist who worked mainly in polymer chemistry. He introduced the representation of polymer-chain lengths by statistical distributions and contributed to the theories of nonlinear polymers and gelation. He was awarded the Nobel prize in Chemistry in 1974 for his fundamental achievements, both theoretical and experimental, in the physical chemistry of the macromolecules. *See* Reference [110].
Flory-Huggins Interaction Parameter	A parameter in Flory-Huggins theory that measures the change upon mixing of the energy of interaction between a solvent molecule and a polymer segment, measured in units of RT.
Flory-Huggins Theory	A description of polymer solutions in which the polymer molecules adopt random coil configurations. The polymer groups or solvent molecules are taken to occupy positions in a lattice. This arrangement allows calculation of the thermodynamics of polymer mixing in solutions.
Flory-Krigbaum Theory	A description of nonideal polymer solutions; it statistically describes what happens as polymer coils approach each other. This description of steric stabilization has also been applied to other dispersions, such as suspensions.
Flory Point	$\rightarrow$ Theta Temperature.
Flory Temperature	$\rightarrow$ Theta Temperature.
Flotation	$\rightarrow$ Froth Flotation, $\rightarrow$ Sedimentation.
Flotation De-Inking	$\rightarrow$ De-Inking.
Flotation-Washing De-Inking	$\rightarrow$ De-Inking.
Flowables	$\rightarrow$ Suspension Concentrates.
Flow Improver	$\rightarrow$ Drag Reducing Agent.
Flow-Line Treating	In oil production or processing, the process in which emulsion is continuously broken and separated into oil and water bulk phases. This process is an alternative to batch treating of emulsions. $\rightarrow$ Treater.
Fluid Film	A thin-fluid phase, usually of thickness less than about 1 µm. Such films can be specified by abbreviations similar to those used for emulsions; for example, some common designations are: A/W/A for a water film in air W/O/W for an oil film in water O/W/O for a water film in oil W/O/A for an oil film between water and air.

Fluid films are usually unstable to breakage caused by rupture: thinning to the point of allowing contact of the separating phase(s). There can, however, be film thicknesses at which a film is stable or metastable to thickness changes. Films with this property are equilibrium films. Otherwise fluid films can be distinguished by rapid (mobile film) or slow (rigid film) thickness changes. → Black Film.

Fluidifier   → Plasticizer.

Fluidity   The inverse of viscosity.

Fluidized-Bed Processing   A coating or fabrication method in which particles suspended in a fluidized bed are coated with another material.

Fluidizer   → Plasticizer.

Fluorescence Microscopy   Light microscopy in which ultraviolet light induces fluorescence in a specimen. The fluorescent light then forms the magnified image, in either transmitted- or reflected-light modes (the ultraviolet light is filtered-out at this stage).

Fluorescence Photoactivation Localization Microscopy   (F-PALM) → Super-Resolution Microscopy.

Fluorescent Whitening Agents   (FWA) → Optical Brighteners.

Fluorochroming   The use of fluorescent dye(s) to stain a specimen and make it visible for microscopic study.

Fluoroprotein Foam   (FP) A fire-extinguishing foam based on hydrolyzed protein and perfluorinated surfactants. Used, for example, on storage tanks. → Film Forming Fluoroprotein Foam, → Aqueous Film Forming Foam, → Alcohol Resisting Aqueous Film Forming Foam, → Fire Extinguishing Foam.

Flux   The flow rate of matter or energy per unit area.

Fly Knives   The rotating cutting blades in a cutting mill machine for comminution.

Foam   A dispersion of gas bubbles in a liquid, in which at least one dimension falls within the colloidal size range. Thus a foam typically contains either very small bubble sizes or, more commonly, quite large gas bubbles separated by thin liquid films. The thin liquid films are called "lamellae" (or "laminae"). Sometimes distinctions are drawn as follows. Concentrated foams, in which liquid films are thinner than the bubble sizes and the gas bubbles are polyhedral, are termed "polyederschaum". Low-concentration foams, in which the liquid films have thicknesses on the same scale or larger than the bubble sizes and the bubbles are approximately spherical, are termed "gas emulsions", "gas dispersions", or "kugelschaum". → Evanescent Foam, → Froth, → Aerated Emulsion, → Solid Foam.

Foam Blanket   Several kinds of foams have been developed for use as temporary blankets to suppress odorous emissions (in sanitary landfill sites), hazardous emissions (fire-fighting and chemical/biological weapon containment), dusts (mine ore and tailings piles), noise- and pressure waves (blasting), and to repel pests (from plants and buildings). *See also* Reference [28].

Foam Booster   → Foaming Agent.

Foam Breaker   Any agent that acts to reduce or eliminate foam stability. Also termed "defoamer". A more general term is "antifoaming agent". → Antifoaming Agent.

Foam Drainage   The drainage of liquid from liquid lamellae separating bubbles in a foam. → Fluid Film.

Foam Drilling Fluid   A drilling fluid comprising air, water, and a foaming agent (surfactant). These substances travel into a well as a mist, then change into a foam before returning up the annulus. → Air Drilling Fluid, → Stable Foam, → Stiff Foam.

Foam Emulsion   → Aerated Emulsion.

Foam Film   → Liquid Film.

Foamer   → Foaming Agent.

Foam Flooding   (1) Enhanced oil recovery: The process in which a foam is made to flow through an underground reservoir. The foam, which can either be generated on the surface and injected or generated *in situ*, is used to increase the drive fluid viscosity and improve its sweep efficiency.
(2) Petroleum processing: In refinery distillation and fractionation towers, the occurrence of foams, which can carry liquid into regions of the towers intended for vapour.

Foam Fractionation   A separation method in which a component of a liquid preferentially adsorbed at the liquid-gas interface is removed by foaming the liquid and collecting the foam produced. Foaming surfactants can be separated in this manner.

Foaminess   A measure of the persistence of a foam (the time an average bubble exists before bursting). Ideally independent of the apparatus and procedure used, and characteristic of the foaming solution being tested. In practice these ideals have not been achieved but some approaches to determining foaminess using dynamic foam stability tests have been reviewed by Bikerman [71]. → Dynamic Foam Test.

Foaming Agent	Any agent that acts to stabilize a foam. The foaming agent can make it easier to form a foam or provide stability against coalescence. Foaming agents are usually surfactants. Also termed "foam booster", "whipping agent", and "aerating agent".
Foaming Power	→ Increase of Volume upon Foaming.
Foam Inhibitor	Any agent that acts to prevent foaming. Also termed "foam preventative". A more general term is "antifoaming agent". → Antifoaming Agent.
Foam Number	A relative drainage rate test in which a foam is formed in a vessel and thereafter the remaining foam volume is determined as a function of time. The foam number is the volume of bulk liquid that has separated after a specified interval, expressed as a percentage of the original volume of liquid foamed.
Foamover	In an industrial process vessel, unwanted foam can occasionally build up to such an extent that it becomes carried out the top of the vessel ("foamover") and on to the next part of the process. This carry-over of foam and any entrained material that comes with it is frequently detrimental to other parts of a processing operation.
Foam Preventative	→ Foam Inhibitor, → Antifoaming Agent.
Foam Quality	The gas volume fraction in a foam. Expressed as a percentage this fraction is sometimes referred to as "Mitchell foam quality". In three-phase systems other measures are used. For example, when foams are formulated to contain solid particles as well, the slurry quality, Q_s, which gives the volume fraction of gas plus solid, can be used: $Q_s = (V_g + V_s)/(V_g + V_s + V_1)$, where V_g, V_s, and V_1 denote the volumes of gas, solid, and liquid phases, respectively.
Foam Stability	→ Foaminess.
Foam-Stimulation Fluid	A foam, aqueous or nonaqueous, that is injected into a petroleum reservoir to improve the productivity of oil- or gas-producing wells. Some mechanisms of action for foam- stimulation fluids include fracturing, acidizing to increase permeability, and diversion of flow.
Foam Texture	The bubble size distribution in a foam. For foams in porous media, it can be expressed in terms of the length scale of foam bubbles as compared with that for the spaces confining the foam. When the length scale of the confining space is comparable to or less than the length scale of the foam bubbles, the foam is sometimes termed

	"lamellar foam", to distinguish it from the opposite case, termed "bulk foam".
Focused Beam Reflectance Measurement	(FBRM) A particle sizing technique in which a laser beam is projected into a dispersion of particles, droplets, or bubbles through a window and focused to a small spot close to the window. The spot is made to rotate at high speed. As a dispersed species passes by the window, the focused beam passes one edge of the species and some of the light is backscattered until the beam passes the opposite edge. From this can be determined a chord length between two points on the edge of the species. With sufficient particles, at random orientations to the path of the rotating spot, an average chord length can be determined, yielding a measure of average species size. A chord length distribution can also be determined, yielding a measure of species size distribution.
Fog	A dispersion of a liquid in a gas. In some definitions fog is characterized by a particular droplet size range, while in others fog refers to mist having a high enough droplet concentration to obscure visibility. → Aerosol of Liquid Droplets, → Mist, and *See* Table 5.
Food Colloid	Many foods are composed of colloidally dispersed phases. The colloidal properties have a great bearing on their texture, appearance, and stability against separation. Examples: suspensions (chocolate), emulsions (milk), foams (ice-cream), and gels (mayonnaise). *See* Reference [111].
Ford-Cup Viscosity	A parameter intended to approximate fluid viscosity. It is measured by an orifice viscometer constructed according to an industry-standard design. The measurement unit is Ford seconds.
Foreign Colloid	→ Pseudocolloid.
Forward Scattering	→ Light Scattering.
Fowkes Equation	A means for estimating the interfacial tension between two liquids, based on the surface tensions and molecular properties of each liquid. *See* Table 8.
Fowler-Guggenheim Equation	An extension of the Langmuir isotherm equation that allows for lateral interactions among adsorbed molecules.
FP	→ Fluoroprotein Foam.
F-PALM	Fluorescence Photoactivation Localization Microscopy. → Super-Resolution Microscopy.
Fractal	A structure that has an irregular shape under all scales of measurement. The fractal dimension of a species is the exponent D to

which a characteristic dimension must be raised to obtain proportionality with the overall size of the species. Fractal dimensions are used for species having a dimensionality between 2 and 3, such as many particle aggregates.

Fractal Dimension	→ Fractal.
Fradkina Equation	For predicting the relative permittivity of emulsions. *See* Table 9.
Fraunhofer, Joseph von (1787–1826)	A German instrument maker and physicist known for describing atomic absorption bands (dark lines termed Fraunhofer lines) in spectra of the sun and stars. He built many kinds of optical instruments, including the spectroscope, and pioneered the use of a diffraction grating to produce a spectrum from white light.
Free-Draining Polymer	Polymer molecules in their extended rather than coiled configuration, so that laminar flow of solvent can occur along the molecules.
Free Energy	A measure of the balance of energetic and entropic forces in a system. For systems maintained at constant pressure, the free energy is referred to as the "Gibbs free energy" (now frequently termed "Gibbs energy"); $G = H - TS$, where H is the enthalpy, T is temperature, and S is entropy. For systems maintained at constant volume, the free energy is referred to as the "Helmholtz free energy" ("Helmholtz energy"); $A = E - TS$, where $E = H - PV$ (P is pressure and V is volume).
Free Molecules	In polymer- or surfactant-containing systems, the molecules of polymer or surfactant dissolved in solution: that is, those that are not adsorbed or precipitated. For surfactant solutions, free surfactant includes those molecules present in micelles.
Free Polymer	→ Free Molecules.
Free Surfactant	→ Free Molecules.
Free Volume	The volume in a system, or near an interface, that is available and not occupied by other molecules or dispersed species. → Excluded Volume.
Free Water	The readily separated, nonemulsified water that is coproduced with oil from a production well, or other practical process.
Free-Water Knockout	(FWKO) A vessel designed to separate readily separated (nonemulsified or "free") water from oil or an oil-containing emulsion. Further water and solids removal can be accomplished in a treater.
Freeze-Fracture Method	A sample preparation technique used in electron microscopy in which specimens are quickly frozen in a cryogen, then cleaved to expose interior surfaces. In some techniques the sample is then

observed directly in an electron microscope equipped with a cryogenic stage; in other cases the cleaved sample is coated with a metal coating to produce a replica, which is observed in the electron microscope. → Scanning Electron Microscopy, → Transmission Electron Microscopy, → Replica.

Freezing Front Method → Solidification Front Method.

Fremdkolloide → Pseudocolloid.

Frenkel-Halsey-Hill Isotherm (FHH Isotherm). An adsorption isotherm equation that accounts for the possibility that adsorbates may possess a distribution of adsorption energies even in the first layer of adsorbed gas, as a monolayer is approached, and in multilayer adsorption the adsorption energy decreases towards the heat of liquefaction (rather than in a step function after the first layer as is assumed in the BET model). The FHH isotherm tends to fit adsorption data over a wider range of relative pressures then the BET isotherm. → Adsorption Isotherm.

Freundlich, Herbert Max Finlay (1880–1941) A German colloid and interface scientist known for contributions in flocculation, electrokinetics and rheology (he coined the terms "thixotropy" and "rheopexy"). He worked in fundamental research as well as industrial process research. The Freundlich equation (adsorption) is commonly attributed to him although he apparently neither originated nor supported it. Freundlich wrote a textbook in the field, the third edition of which was translated into English (Colloid and Capillary Chemistry). *See* References [112, 113].

Freundlich Isotherm An empirical adsorption isotherm equation for heterogeneous surfaces based on exponentially decreasing enthalpy of adsorption with increasing surface coverage. The amount adsorbed per mass of adsorbent increases with increasing equilibrium solute concentration at all concentrations and does not achieve monolayer coverage. → Adsorption Isotherm.

Fricke Equation An equation for predicting the conductivity of dispersions. *See* Table 10.

Friction The surface resistance of a body to motion. The coefficient of friction is given by the frictional force needed to move one surface over another divided by the load normal to the direction of motion along the surfaces (Amonton's law). The static coefficient of friction is that involving the force needed to initiate motion, and the kinetic coefficient of friction involves the force needed to maintain a given rate of motion. The science of friction and lubrication is known as tribology.

Friction Factor	In the rheology of a dispersion, the friction factor relates to the dissipation of energy due to friction at the surfaces of the dispersed species (i.e., due to drag).
Friction Force Microscopy	(FFM) Atomic force microscopy in which the cantilever tip remains in constant contact with the sample surface while being pulled over it at constant velocity. Both topographical information and torsional (friction) forces are measured along the scanning direction and used in micro- and nanoscale studies of friction and lubrication (micro- and nanotribology). Also termed contact-mode atomic force microscopy (C-AFM), scanning force microscopy, or lateral force microscopy. → Scanning Probe Microscopy.
Froth	A type of foam in which solid particles are also dispersed in the liquid (in addition to the gas bubbles), as in froth flotation. The solid particles can even be the stabilizing agent; alternatively, the foam layer produced at the top of a separation vessel or distillation tower. The term sometimes refers simply to a concentrated foam, but this usage is not preferred.
Frother	→ Frothing Agent.
Froth Flotation	A separation process utilizing flotation, in which particulate matter becomes attached to gas (foam) bubbles. The flotation process produces a product layer of concentrated particles in foam termed froth. Variations include dissolved-gas flotation, in which gas is dissolved in water that is added to a colloidal dispersion. As microbubbles come out of solution they attach to the colloidal species and cause them to float. → Carrier Flotation, → Emulsion Flotation, → Floc Flotation, → Roughing Flotation, → Scalping Flotation, → Scavenging Flotation, → Slime Coating, → One-Second Criterion.
Frothing Agent	Any agent that acts to stabilize a froth. Can make it easier to form a froth and provide stability against coalescence. Frothing agents are usually surfactants. Example: cresol. Also called frother; analogous to foaming agent. → Collector; → Froth Flotation, → Modifier.
Fuchs Surface Area	The portion of the surface area of an aerosol particle that interacts with a carrier gas. Also termed active surface area. → Aerosol Diffusion Charging, → Epiphaniometer.
Fuerstenau, Douglas W. (1928–)	An American metallurgical engineer known for his contributions (over 400 publications) to mineral processing, covering virtually every aspect of this field, including comminution, agglomeration, mineral flotation, and fine particle processing. See References [114, 115].

Fullerene	A closed-cage structure, such as a hollow sphere or ellipsoid, having more than twenty carbon atoms and consisting of three-coordinate carbon atoms. Buckminsterfullerene {$(C_{60}$-$I_h)$[5, 6]fullerene} is the most common naturally occurring fullerene (it occurs in soot), and is also termed Buckyball. Carbon nanotubes are cylindrical fullerenes.
Fulvic Acids	$\rightarrow$ Humic Substances.
Fume	An aerosol of solid particles (dispersion of solid particles in gas) in which the particle sizes are less than 1 μm in diameter. Fumes can arise from the condensation of vapors from a chemical or physical reaction. The term fume is also sometimes used to denote the simultaneous presence of a gas in smoke. $\rightarrow$ Aerosol.
Fumed Powder	A powder that has been collected from a fume.
Fumed Silica	A powdered form of silicon dioxide generated by thermal pyrolysis.
Furcellaran	A water-soluble mixture of sulfated galactans derived from seaweeds. Also termed "Danish agar". Furcellaran sols are similar to those of agar and can be quite viscous; can readily form gels; and can be used to stabilize certain suspensions, foams, and emulsions. Furcellaran is used in many applications, especially in foods and medicines. *See also* Reference [35], $\rightarrow$ Seaweed Colloids.
Furnace-Flow Processing	A fine particle preparation method in which particles are synthesized from a saturated vapour. This method can be used to prepare colloidal and/or nanoparticles.
FWA	Fluorescent whitening agents. $\rightarrow$ Optical Brighteners.
FWKO	$\rightarrow$ Free-Water Knockout.

G

Galvani Potential	$\rightarrow$ Inner Potential.
Gangue	In mineral processing the gangue is the unwanted portion of a raw mineral ore. In classical froth flotation, for example, the feed ore is separated into hydrophobic (valuable) particles that are floated, skimmed off in a froth, and collected. The hydrophilic (unwanted) particles that are not floated represent the gangue.
Gas Aphrons	$\rightarrow$ Microgas Emulsions.
Gas Black	$\rightarrow$ Carbon Black.
Gas Dispersion	$\rightarrow$ Foam, $\rightarrow$ Solid Foam, $\rightarrow$ Gas Emulsion.
Gas Emulsion	"Wet" foams in which the liquid lamellae have thicknesses on the same scale or larger than the bubble sizes. Typically in these cases the gas bubbles have spherical rather than polyhedral shape. Other synonyms include "gas dispersion" and "kugelschaum". If the bubbles are very small and have a significant lifetime, the term "microfoam" is sometimes used. In petroleum production the term is used to specify crude oil that contains a small volume fraction of dispersed gas. $\rightarrow$ Foam.
Gas-Phase Synthesis Method	Any of several fine particle preparation methods in which one or more of the feedstock materials goes through a gas phase, nucleation, and then growth through condensation, coagulation, and/or coalescence. Examples include physical vapour deposition, chemical vapour deposition, furnace-flow processing, laser pyrolysis gas-phase synthesis, and plasma processing. These methods can be used to prepare colloidal and/or nanoparticles.
GD-AAS	Glow Discharge Atomic Absorption Spectrometry. $\rightarrow$ Glow Discharge Spectrometry.

Dictionary of Nanotechnology, Colloid and Interface Science. Laurier L. Schramm
Copyright © 2008 WILEY-VCH Verlag GmbH & Co. KGaA, Weinheim
ISBN: 978-3-527-32203-9

GD-AFS	Glow Discharge Atomic Fluorescence Spectrometry. $\rightarrow$ Glow Discharge Spectrometry.
GDS	$\rightarrow$ Glow Discharge Spectrometry.
GDMS	Glow Discharge Mass Spectrometry. $\rightarrow$ Glow Discharge Spectrometry.
GD-OES	Glow Discharge Optical Emission Spectrometry. $\rightarrow$ Glow Discharge Spectrometry.
Gegenion	$\rightarrow$ Counterions.
Gel	A dispersion [4, 6, 116–118] in the form of a suspension or polymer solution that behaves as an elastic solid or semi-solid rather than a liquid. A dried-out gel is termed a xerogel. Examples: gels of gelatin solutions or of clay suspensions. Many other uses of this term and related terms are described in Reference [119].
Gelatin	$\rightarrow$ Protein Colloid.
Gel Filtration Chromatography	$\rightarrow$ Chromatography.
Gel Foam	A foam that, in addition to the stabilizing surfactants, contains polymer and a cross-linking agent. The foam is first generated as a polymer-thickened foam, and after a delay period, gels. $\rightarrow$ Stiff Foam.
Gellan Gum	A high molecular weight polysaccharide gum. It is used as a thickening or gelling agent, particularly in foods such as bakery fillings and frostings, dairy products, jams, puddings, and sauces. It is also used in personal care products, such as lotions and creams, hair care products, and toothpaste. $\rightarrow$ Seaweed Colloids, $\rightarrow$ Hydrocolloid.
Gel Permeation Chromatography	$\rightarrow$ Chromatography.
Gel Point	The onset of gelation, as denoted by the stage at which a liquid or dispersion begins to exhibit an increased viscosity and elastic properties. $\rightarrow$ Gel.
Gemini Surfactant	A type of dimeric surfactant molecule having two long hydrocarbon (tail) chains and two polar head groups linked together by a short chain. The connecting chain may be attached to the two head groups or to the two tails. Gemini surfactants usually have different physical properties, such as different micelle structures, from single-chain surfactants having the same number of carbon atoms per head group.

Generalized Plastic Fluid	A fluid characterized by both of the following: the existence of a finite shear stress that must be applied before flow begins (yield stress), and pseudoplastic flow at higher shear stresses. → Bingham Plastic Fluid.
Germs	(Nucleation) → Nuclei.
g-Forces	→ Relative Centrifugal Force.
Giant Nuclei	Aerosol particles having diameters greater than 2.0 micrometres. Example: large pollens. *See also* Table 6, → Aerosol of Solid Particles.
Gibbs Adsorption	→ Gibbs Surface.
Gibbs Adsorption Isotherm	→ Gibbs Isotherm.
Gibbs Dividing Surface	→ Gibbs Surface.
Gibbs-Duhem Equation	The thermodynamic relationship between the partial molar properties of one component in a system and the partial molar properties of other components as a function of composition.
Gibbs Effect	The decrease in surface or interfacial tension that occurs as surfactant concentration increases towards the critical micelle concentration.
Gibbs Elasticity	→ Film Elasticity.
Gibbs Energy	Gibbs free energy. → Free Energy.
Gibbs Energy of Attraction	When two dispersed-phase species approach, they can attract each other as a result of such forces as the London-van der Waals forces. The Gibbs energy of attraction can be thought of as the difference between Gibbs attractive energies of the system at a specified separation distance and at infinite separation. Although IUPAC [4] has discouraged the use of the synonyms "potential energy of attraction" and "attractive potential energy", they are still in common usage. → Gibbs Energy of Interaction, → Gibbs Energy of Repulsion.
Gibbs Energy of Interaction	When two dispersed-phase species approach, they experience repulsive and attractive forces such as electrostatic repulsion and van der Waals attraction. The Gibbs energy of interaction can be thought of as the difference between Gibbs energies of the system at a specified separation distance and at infinite separation. An example of the dependence of Gibbs energy of interaction and distance of separation is that calculated from DLVO theory. Although IUPAC [4] has discouraged the use of the synonyms "potential energy of interaction" and "total potential energy of interaction", they are still in common usage. → Gibbs Energy of Attraction, → Gibbs Energy of Repulsion, → DLVO Theory, → Primary Minimum, → Short-Range Interaction Forces.

Gibbs Energy of Repulsion	When two dispersed-phase species approach, they can repel each other as a result of such forces as electrostatic repulsion. The Gibbs energy of repulsion can be thought of as the difference between Gibbs repulsive energies of the system at a specified separation distance and at infinite separation. Although IUPAC [4] has discouraged the use of the synonyms "potential energy of repulsion" and "repulsive potential energy", they are still in common usage. → Gibbs Energy of Attraction, → Gibbs Energy of Interaction, → Short-Range Interaction Forces.
Gibbs Film Elasticity	→ Film Elasticity.
Gibbs Free Energy	Now frequently termed "Gibbs energy". → Free Energy.
Gibbs Isotherm	An equation that relates the Gibbs surface excess (amount adsorbed) to the change in interfacial tension with activity of the adsorbing species.
Gibbs, Josiah Willard (1839–1903)	An American physicist who became famous for his contributions to thermodynamics and mathematical physics. Eponyms include "Gibbs free energy", "Gibbs phase rule", "Gibbs-Dalton law", "Gibbs-Duhem formula", and the "Gibbs-Helmholtz equation" (all in thermodynamics). *See* References [120, 121].
Gibbs-Marangoni Effect	The effect in thin liquid films and foams whereby stretching an interface causes the surface excess surfactant concentration to decrease, hence surface tension to increase (Gibbs effect); the surface tension gradient thus created causes liquid to flow toward the stretched region, thus providing both a "healing" force and a resisting force against further thinning (Marangoni effect). Sometimes referred to simply as the "Marangoni effect".
Gibbs Phase Rule	→ Phase Rule.
Gibbs-Plateau Border	→ Plateau Border.
Gibbs Ring	→ Plateau Border.
Gibbs Surface	A geometrical surface chosen parallel to the interface and used to define surface excess properties such as the extent of adsorption. → Gibbs Surface Excess.
Gibbs Surface Concentration	The Gibbs surface excess adsorption amount divided by the area of the interface.
Gibbs Surface Elasticity	→ Film Elasticity.
Gibbs Surface Excess	The excess amount of a component present in a system over that present in a reference system of the same volume as the real system, and in which the bulk concentrations in the two phases remain uniform up to the Gibbs dividing surface. The terms

"surface excess concentration" or "surface excess" have now replaced the earlier term "superficial density".

Girifalco-Good Equation | An equation for estimating interfacial tension based on the surface tensions and molecular volumes of the two liquids. *See* Table 8.

Girifalco-Good-Fowkes-Young Equation | An equation relating contact angle through a liquid, between a gas and a solid, to the components of surface tension of the solid and liquid that are due to dispersion (London-van der Waals) forces and the surface tension of the liquid.

Glass-Transition Temperature | The temperature at which a polymer changes from a viscous or elastic state into a nonelastic solid state.

Glow Discharge | A form of plasma comprising approximately equal concentrations of positive and negative charges plus a large number of neutral species. An electric field gradient is applied across electrodes in a cell containing an inert or molecular gas at low pressure. As a result, the gas becomes ionized creating free electrons and positively charged ions. The positively charged ions accelerate towards, and collide with, the cathode causing secondary electron emission. The secondary electrons then collide with the gas atoms or molecules causing excitation and ionization. The excitation collisions and subsequent decay cause the light emission, or glow. The ionization collisions create new electrons and ions. The multiplication of the electron-ion processes creates a self-sustaining plasma. → Glow Discharge Spectrometry.

Glow Discharge Atomic Absorption Spectrometry | (GD-AAS) → Glow Discharge Spectrometry.

Glow Discharge Atomic Fluorescence Spectrometry | (GD-AFS) → Glow Discharge Spectrometry.

Glow Discharge Mass Spectrometry | (GDMS) → Glow Discharge Spectrometry.

Glow Discharge Optical Emission Spectrometry | (GD-OES) → Glow Discharge Spectrometry.

Glow Discharge Spectrometry | (GDS) An analytical method in which a sample to be studied is used as the cathode of a glow discharge (a kind of plasma). The sample is bombarded by positive ions causing the release of secondary electrons and atoms of the sample material (sputtering). The sputtered atoms become ionized due to collisions with electrons and ions in the glow discharge plasma. These sample ions can be studied with a mass spectrometer, hence the term Glow Discharge Mass Spectrometry (GDMS). Collisions between the sample ions and the gas atoms or molecules cause excitation and subsequent decay, which in turn causes characteristic photon emissions. The latter can be studied with an optical emission spectrometer, hence the term Glow Discharge Optical Emission Spectrometry (GD-OES). With the use of an external light source, the sputtered atoms can also be studied by atomic absorption and fluorescence techniques, hence the terms Glow Discharge Atomic Absorption Spectrometry (GD-AAS), and Glow Discharge Atomic Fluorescence Spectrometry (GD-AFS). → Glow Discharge.

Glue	→ Adhesive, → Protein Colloid.
Gold Number	A test used to provide a basis for comparing the stabilizing, or "protecting," ability of polyelectrolytes in suspensions. In this test [26] the "gold number" is defined as the number of milligrams of polyelectrolyte that just prevent the flocculation of 10 ml of a gold sol by 1 ml of a 10% solution of sodium chloride. The smaller the gold number the greater the protecting, or stabilizing, power of the polyelectrolyte.
Gouy-Chapman Layer	→ Electric Double Layer.
Gouy-Chapman Theory	A description of the electric double layer in a colloidal dispersion in which one layer of charge is assumed to exist as a uniform charge distribution over a surface, and the counterions are treated as point charges distributed throughout the continuous, dielectric phase.
Gouy Layer	→ Electric Double Layer.
Gouy, Louis Georges (1854–1926)	A French physicist who investigated problems in physics and in colloids, including studies on Brownian motion. He is perhaps best known to colloid science for his theory of the diffuse double layer at a charged interface. Related eponyms include the "Gouy layer" and "Gouy-Chapman" and "Gouy-Chapman-Stern theories". See Reference [122].
Graham, Thomas (1805–1869)	A Scottish physical chemist and the founder of colloid chemistry, Graham is known for contributions to diffusion phenomena in liquids leading to the establishment of colloids and the colloidal state; he coined the terms "colloid" and "dialysis". Graham is also known for contributions to under standing the diffusion of gases, including the law that the velocities of gases are inversely proportional to the square roots of their densities. He was the first president of the Chemical Society of London. See References [123, 124].
Granular Media Filtration	→ Filtration.
Granulation	The process in which particles become aggregated into larger grains, or granules.

Granule	Sometimes used to describe particles having sizes greater than about 2000 μm, depending on the classification system used. Also called gravel. *See* Table 12.
Graphite	A naturally occurring, soft (low hardness) form of carbon. Graphite is used in the manufacture of lead pencils, lubricants, and paints. Also termed black lead, or plumbago. Powdered graphite is sometimes referred to as blacking.
Graphene	A two-dimensional sheet of carbon. → Carbon Nanotube.
Gravel	→ Granule.
Gravity Separator	→ Separator.
Grease	Greases, or metalworking fluids, provide lubricity and cooling. They also have to exhibit a significant yield stress since they are intended to flow into tight spaces when injected under pressure and then remain in place. These fluids are usually non-aqueous suspensions. Example: a gelled solution of lithium 12-hydroxy stearate surfactant in mineral oil. Some more modern semi-synthetic metalworking oils are microemulsions.
gs	→ Relative Centrifugal Force.
Guerbet Alcohol	Guerbet alcohols are 2-alkyl alkanols in which the carbon numbers can typically be in the range C_{12} to C_{24}. Such compounds are produced from linear fatty alcohols, by the Guerbet reaction, and have lower melting points than their corresponding saturated fatty alcohols. Guerbet acids, such as carboxylic acids, are derived from the corresponding alcohols. *See* Reference [125].
Guinier Plot	A way of plotting scattering data (static light scattering, small-angle neutron scattering or small-angle X-ray scattering), essentially in terms of intensity versus angle, for particle dispersions. The plot can be used to determine molar mass and the radius of gyration. This can be compared with a Zimm plot for polymer solutions.
Gum	Any hydrophilic plant material, or derivative, that forms a viscous dispersion or solution with water. Example: gum arabic (acacia gum) is derived from Acacia trees and is used in paints, inks, adhesives, and textiles.
Gum Arabic	→ Gum.
Gun Barrel	A type of settling vessel used to separate water and oil from an emulsion. Typically, a heated emulsion is treated with demulsifier and introduced into the gun barrel where water settles out and is drawn off. Any produced gas is also drawn off.
Guth-Gold-Simha Equation	An empirical equation for estimating the viscosity of a dispersion. *See* Reference [126] and *see* Table 11.

H

Hagen–Poiseuille Law	Poiseuille flow is the steady flow of incompressible fluid parallel to the axis of a circular pipe or capillary. The Hagen–Poiseuille law is an expression for the flow rate of a liquid in such tubes. It forms the basis for the measurement of viscosities by capillary viscometry.
Half-Colloid	→ Lyophilic Colloid.
Half-Micelle	→ Hemimicelle.
Halo	In particle characterization, the false perimeter apparent in a particle image.
Hamaker Constant	In the description of the London–van der Waals attractive energy between two dispersed bodies, such as particles or droplets, the Hamaker constant is a proportionality constant characteristic of the particle composition. It depends on the internal atomic packing and polarizability of the droplets. Also termed the "van der Waals–Hamaker constant".
Hamaker, Hugo Christiaan (1905–)	A Dutch physicist known for his contributions to the understanding of flocculation. Hamaker derived an expression for the London–van der Waals interaction energy consisting of a geometric factor (dependent on the size, shape and separation of the bodies involved) and a factor dependent only on the material properties (hence the Hamaker constant). Hamaker also worked in physics (high voltage electrodeposition from suspensions in nonaqueous solvents) and statistics (applied to quality control in factories). *See* Reference [127].
Hammer Mill	A device for reducing the particle size of a solid, (for example, a pigment), that uses centrifugal force to drive the solid between rotating "hammers" and a stationary ring-shaped "anvil".

Dictionary of Nanotechnology, Colloid and Interface Science. Laurier L. Schramm
Copyright © 2008 WILEY-VCH Verlag GmbH & Co. KGaA, Weinheim
ISBN: 978-3-527-32203-9

Hanai Equation	An equation for predicting the conductivities or relative permittivities of dispersions. *See* Tables 9 and 10.
Hanging Bubble Method	→ Pendant Drop Method.
Hanging Drop Method	→ Pendant Drop Method.
Hardness	(1) Mineralogy. The resistance of a material to deformation, particularly resistance to scratches or indentation. In mineralogy scratch hardness is judged by comparison with a set of (usually ten) reference materials. The Mohs scale of scratch hardness ranges from a low value of 1 (talc) to a high value of 10 (diamond). *See* Table 17. (2) Metallurgy. In metallurgy indentation hardness is denoted by the pressure required to create an indentation in a material. Macrohardness generally involves large samples and large indentations. Microhardness involves small indentations requiring a microscope for their observation. Nanohardness involves nanoscale indentation (nanoindentation) and nanoscale indentation forces (∼1 nN). Materials with the greatest pressure-resistance to indentation (>40 GPa) have what is termed superhardness. Examples include some composites (∼50–80 GPa) and diamond (∼70–100 GPa).
Harkins–Jura Isotherm	An adsorption isotherm equation that accounts for the possibility of multilayer adsorption. → Adsorption Isotherm.
Harkins, William Draper (1873–1951)	An American physical chemist known to colloid and interface scientists for his work in experimental physical chemistry, especially surface chemistry, including surface and interfacial tensions, adsorption and monomolecular films. Also known for his demonstration of the poisonous effects of smelter pollution in the early 1900's and for his work in nuclear physics, on the structure and reactions of atoms. Eponyms include the "Harkins–Wilson theory" (atom building) and the "Harkins–Jura equation" (an adsorption isotherm equation). *See* References [128, 129].
Hatschek Equation	An equation for predicting the viscosity of emulsions. *See* Table 11.
Haze	An aerosol of solid particles (dispersion of solid particles in gas) in which the particle sizes are smaller than can be seen without the aid of a microscope. Hazes may contain water droplets as well. → Aerosol, → Fume, → Dust.
HDAs	→ High-Dispersed Aerosols.
HDC	→ Hydrodynamic Chromatography.
Head Group	The lyophilic functional group in a surfactant molecule; in aqueous systems, the polar group of a surfactant. → Surfactant, → Surfactant Tail.
Heater Treater	→ Treater.

Heavy Crude Oil	A naturally occurring hydrocarbon having a viscosity less than 10,000 mPa s at ambient deposit temperature, and a density between 934 and 1000 kg/m^3 at 15.6 °C. *See* References [50–52].
HEED	→ High-Energy Electron Diffraction.
Hegman Gauge	A method for quickly determining the approximate size of the largest particles in a suspension. The gauge comprises a plate with a precision-machined channel that is tapered to provide slight through deep depression. A sample of suspension is placed in the channel and levelled with a flat blade. The channel is then examined and the point at which coarse particles begin to appear is identified. An engraved scale shows the depth.
HEIS	High-Energy Ion Scattering. → Ion-Scattering Spectroscopy.
Helmholtz Condenser	→ Helmholtz Double Layer.
Helmholtz Double Layer	A simplistic description of the electric double layer as a condenser (the Helmholtz condenser) in which the condenser plate separation distance is the Debye length. The Helmholtz layer is divided into an inner Helmholtz plane (IHP) of adsorbed, dehydrated ions immediately next to a surface, and an outer Helmholtz plane (OHP) at the center of a next layer of hydrated, adsorbed ions just inside the imaginary boundary where the diffuse double layer begins. That is, both Helmholtz planes are within the Stern layer.
Helmholtz Energy	Helmholtz free energy. → Free Energy.
Helmholtz Free Energy	Now frequently termed "Helmholtz energy". → Free Energy.
Helmholtz, Hermann von (Ludwig Ferdinand) (1821–1894)	A German physician and scientist known for his contributions to mathematics and many of the sciences. He is known to colloid and interface scientists for his co-discovery of the second law of thermodynamics and for contributions in electric double layer theory, not to mention many other broad areas, especially physiology. Eponyms include the "Gibbs–Helmholtz equation", "Helmholtz free energy", and the "Helmholtz double layer (Helmholtz–Perrin theory)".
Helmholtz Plane	→ Helmholtz Double Layer.
Helmholtz-Smoluchowski Equation	→ Smoluchowski Equation.
Hemacytometer	A type of particle or droplet sizing and counting chamber used in microscopy. The chamber contains an accurately ruled grid of squares and, with the cover slip in place, holds a specified volume of typically 0.1 μL. Originally developed for and usually used for counting blood cells, it is also used for particle or droplet counting and sizing in other suspensions and emulsions.

Hemimicelle	An aggregate of adsorbed surfactant molecules that can form, distinct from monolayer formation, the enhanced adsorption due to hydrophobic interactions between surfactant tails. Hemimicelles (half-micelles) are considered to have the form of surface aggregates, or of a second adsorption layer with reversed orientation, somewhat like a bimolecular film. For bilayer surfactant aggregates, the term "admicelles" has also been used [130, 134]. Admicellar chromatography, adsolubilization, and admicellar catalysis make use of media bearing admicelles. → Solloids.
Henry Equation	A relation expressing the proportionality between electrophoretic mobility and zeta potential for different values of the Debye length and size of the species. → Smoluchowski Equation, → Hückel Equation, → Electrophoresis.
Henry Isotherm	A sorption isotherm describing the linear (with respect to concentration) partitioning of a chemical between phases. Also used to describe adsorption at low surface coverage when a linear approximation is appropriate (in the limit of very low surface coverage).
Herringbone Multi-Wall Nanotube	(h-MWNT) → Carbon Nanotube.
Herschel-Bulkley Equation	A rheological flow model in which shear stress is given as the sum of the power law model plus a yield stress. This model has been applied very widely to practical dispersions, including suspensions, emulsions, and foams. For this and other models *see* References [21, 28]. → Robertson–Stiff Equation.
Hertz, Heinrich Rudolf (1857–1894)	A German physicist known for his research on electromagnetic waves and electricity. He was the discoverer of radio waves (Hertzian waves). The name Hertz (Hz) is now given to the unit of frequency. *See* Reference [131].
Heteroagglomeration	→ Heterocoagulation.
Heterocoagulation	The coagulation of dispersed species of different types or those having different states of surface electric charge. Also termed mutual coagulation, heteroagglomeration, heteroflocculation. → Aggregation.
Heterodisperse	A colloidal dispersion in which the sizes of the dispersed species (droplets, particles, etc.) vary. Subcategories are paucidisperse (few sizes) and polydisperse (many sizes). → Monodisperse, → Number-Average Quantities, → Mass-Average Quantities.
Heteroflocculation	→ Heterocoagulation.
Heterogeneous Nucleation	→ Condensation Methods.
Heterotactic Species or Surface	→ Tacticity.

Heterovalent	A term that refers to species having different valencies, (as opposed to homovalent, referring to species having the same valency). In ion exchange the term refers to adsorbing and desorbing species that have different charges.
High-Dispersed Aerosols	(HDAs) Colloidal dispersions of liquids or solids in a gas for which the dispersed particle or droplet sizes are less than 100 nm and have Knudsen numbers greater than one. For such dispersions the aggregation process can be approximated as being that for large gas molecules. $\rightarrow$ Aerosol, $\rightarrow$ Knudsen Number.
High-Energy Electron Diffraction	(HEED) A diffraction technique in which a high-energy electron beam is used. The electrons have high penetrating power, so a grazing angle of incidence is used so that measuring the diffraction pattern will yield information about surface structure. The inelastically scattered electrons, having lower energy, are stopped by grids, and the elastically scattered electrons, having the original energy level, are used to form the image pattern. Derived techniques include reflection high-energy electron diffraction (RHEED) and scanning high-energy electron diffraction (SHEED). $\rightarrow$ Low-Energy Electron Diffraction, *see* Table 7.
High-Energy Ion Scattering	(HEIS) $\rightarrow$ Ion-Scattering Spectroscopy.
High-Energy Milling	An attrition method in which particle sizes are reduced by high-energy erosion due to friction and wear.
High-Energy Surface	Qualitative categorization indicating that a surface has a relatively high surface free energy (usually for ionic or covalently bonded materials). Example: sodium chloride crystals, glass. In contrast, low energy surfaces are those having relatively low surface free energies (usually for van der Waals bonded materials). $\rightarrow$ Low-Energy Surface.
Higher-Order Tyndall Spectra	(HOTS) The spectra obtained for light scattered at various angles from an illuminated dispersion. $\rightarrow$ Tyndall Scattering.
High-Pressure Homogenizer	$\rightarrow$ Homogenizer.
High-Range Water-Reducer	$\rightarrow$ Plasticizer.
High-Resolution Electron Energy Loss Spectroscopy	(HREELS) $\rightarrow$ Characteristic Energy-Loss Spectroscopy.
Hindered Settling	$\rightarrow$ Stokes Settling.
HIOC	$\rightarrow$ Hydrophobic Ionogenic Organic Compound.
HLB Scale	$\rightarrow$ Hydrophile-Lipophile Balance.
HLB Temperature	$\rightarrow$ Phase Inversion Temperature.
HLD	$\rightarrow$ Surfactant Affinity Difference.

h-MWNT	Herringbone Multi-Wall Nanotube. → Carbon Nanotube.
HOC	→ Hydrophobic Organic Contaminant.
Hofmeister Series	→ Lyotropic Series.
Hollow Polymer Nanostructure	→ Nanocapsule.
Homogeneous Nucleation	→ Condensation Methods.
Homogeneous Nucleation Temperature	The temperature below which the rate of a nucleation process increases rapidly. In practice a narrow range of temperatures represents the transition from very slow to very rapid nucleation. → Condensation Methods.
Homogeneous Nucleus	→ Embryo.
Homogenizer	(1) General. Any machine for preparing colloidal systems by dispersion. Examples: colloid mill, blender, ultrasonic probe. (2) High-Pressure Homogenizer. A device for homogenizing (reducing droplet sizes) in coarse emulsions by passing the emulsion through a narrow valve, slit, or orifice at elevated pressure. Turbulence, cavitation, and velocity gradients cause the drop size reduction. Example: Food emulsions requiring small droplet sizes, such as homogenized milk, are frequently made using high-pressure homogenizers.
Homopolymer	→ Polymer.
Homotactic Species or Surface	→ Tacticity.
Homovalent	Species having the same valencies, (as opposed to heterovalent, referring to species having different valencies). In ion exchange, the term refers to adsorbing and desorbing species that have the same charge.
Hooke's Law	Expression of the relation between the force per area applied to an elastic body and the tensile stress, where Young's modulus is the constant of proportionality between them.
Houwink, Roelof (1869–1945)	A Dutch physical chemist known for his work in macromolecular, or polymer, chemistry. He worked with Herman Mark on the characterization of polymers, including viscosity. They developed a non-linear version of Staudinger's equation for the viscosity of polymers that accounted for the flexibility of these molecules. Their non-linear equation is known as the "Staudinger–Mark–Houwink equation".
HREELS	High-Resolution Electron Energy Loss Spectroscopy. → Characteristic Energy-Loss Spectroscopy.

Hückel Equation	A relation expressing the proportionality between electrophoretic mobility and zeta potential for the limiting case of a species that can be considered small and with a thick electric double layer. → Smoluchowski Equation, → Henry Equation, → Electrophoresis.
Hückel, Erich (Armand Arthur Joseph) (1896–1980)	A German physical chemist who, with Peter Debye, developed the Debye-Hückel theory (conductivity) of strong electrolyte solutions (replacing Arrhenius theory). He is also known for his discoveries relating to the structures of benzene and similar compounds that exhibit aromaticity. *See* Reference [132].
Humic Acids	→ Humic Substances.
Humic Substances	Polyaromatic and polyelectrolytic organic acids of high molecular mass (about 800–4000 g/mol or higher) that occur in natural water bodies, soils, and sediments. Although significantly aromatic, these acids can have an appreciable aliphatic component as well and can be surface-active [25]. Humic substances are operationally divided into humic acids and fulvic acids on the basis of solubility as follows: Humic acids are water- soluble above pH 2 but water-insoluble below pH 2; fulvic acids are water-soluble at all pH levels.
Hydration Forces	→ Short-Range Interaction Forces.
Hydraulic Mining	The use, in mining operations, of a water jet to dislodge and move mineral particles from the original deposit and create a suspension. This has been applied to a range of ores, including placer gold and coal.
Hydraulic Spraying	→ Airless Spraying.
Hydrocolloid	Any of the hydrophilic colloidal materials used (mostly) in food products as emulsifying, thickening, and gelling agents. Occasionally termed aquacolloid. Hydrocolloids are mostly carbohydrate polymers, although some are proteins. Examples: agar, carrageenan, dextran, gelatin, guar gum.
Hydrodynamic Chromatography	(HDC) A method of determining the size distribution of dispersed species. A dispersion is made to flow through a packed bed of small beads. Fractions of effluent contain different components according to their rate of travel through the bed. The method is different from gel permeation chromatography in that here the packed beads are not porous. → Chromatography.
Hydrodynamic Diameter	The effective spherical diameter of a particle in a liquid, determined based on its hydrodynamic properties. Hydrodynamic diameter determined by dynamic light scattering is termed the Z-average mean. → Equivalent Spherical Diameter.

Hydrophil Balance	→ Film Balance.
Hydrophile-Lipophile Balance	(HLB scale) An empirical scale categorizing surfactants in terms of their tendencies to be mostly oil-soluble or water-soluble, hence their tendencies to promote W/O or O/W emulsions, respectively. → Phase Inversion Temperature, → Three-Dimensional HLB System.
Hydrophilic	A qualitative term referring to the water-preferring nature of a species (atom, molecule, droplet, particle). For emulsions, "hydrophilic" usually means that a species exhibits an affinity for the aqueous phase over the oil phase. In this example, hydrophilic has the same meaning as oleophobic, but such is not always the case.
Hydrophilic-Lipophilic Deviation	→ Surfactant Affinity Difference.
Hydrophobe	→ Hydrophobic.
Hydrophobic	A qualitative term referring to the water-avoiding nature of a species (atom, molecule, droplet, particle). For emulsions hydrophobic usually means that a species exhibits an affinity for the oil phase over the aqueous phase. In this example, "hydrophobic" has the same meaning as oleophilic, but such is not always the case. A functional group of a molecule that is not very water-soluble is referred to as a "hydrophobe".
Hydrophobically Associating Polymers	Molecules in which surfactant monomers are incorporated into polymer chains. Also termed "polysoaps". One type comprises copolymers of conventional monomer with surfactant macro-monomer. Hydrophobically associating polymers characteristically exhibit enhanced viscosity, brine stability, and shear stability compared with conventional non-associating polymers of equivalent molecular mass. Example: acrylamide copolymerized with nonylphenoxypoly(etheroxy)ethyl acrylate. *See* Reference [133] and → Critical Aggregation Concentration, → Surfactant Macromonomers. Another type of hydrophobically modified polymer comprises hydrophobic endgroups capping a hydrophilic polymer. These molecules are also termed "endcapped polymers".
Hydrophobic Bonding	The attraction between hydrophobic species in water, which arises from the fact that the solvent-solvent interactions are more favorable than the solvent-solute interactions.
Hydrophobic Effect	The partitioning of a substance from an aqueous phase into (or onto) another phase due to its hydrophobicity. Often characterized by an octanol-water partitioning coefficient. → Solvent-Motivated Sorption.

Hydrophobic Index	An empirical measure of the relative wetting tendency of very small solid particles. In one test method, solid particles of narrow size range are placed on the surfaces of a number of samples of water containing increasing concentrations of alcohol (thus providing a range of solvent surface tensions). The percentage alcohol solution at which the particles just begin to become hydrophilic and sink is taken as the hydrophobic index. The corresponding solvent surface-tension value is taken as the critical surface tension of wetting. The technique is also referred to as the "film-flotation technique" [134]) or "sink-float method". → Critical Surface Tension of Wetting.
Hydrophobic Interaction	→ Hydrophobic Bonding.
Hydrophobic Ionogenic Organic Compound	(HIOC) An organic compound that is capable of ionizing, depending upon the solution pH. Upon ionization the properties of the molecule change and its sorption and subsurface migration (in the environment) vary accordingly.
Hydrophobic Organic Contaminant	(HOC) An organic molecule (usually neutral) that has a relatively low solubility in water. Example: many pesticides.
Hydrosol	A dispersion of very small diameter species in water or in aqueous solution. Occassionally termed aquasol. Dispersions of finely divided oil droplets in aqueous solution are sometimes referred to as "oil hydrosols". → Organosol.
Hydrotrope	Any species that enhances the solubility of another. Example: hydrotropes such as alkyl aryl sulfonates (e.g., toluene sulfonate) are added to detergent formulations to raise the cloud point. → Hydrotropy.
Hydrotropic Agent	→ Hydrotrope.
Hydrotropy	A phenomenon in which the solubility of a substance which is only slightly soluble in water is increased by the addition of a second solute. The latter is termed a hydrotrope, or hydrotropic agent. Example: Hydrotropic agents are used to enhance the solubilities of many poorly water-soluble drugs.
Hyperfiltration	→ Filtration.

I

IBA	→ Ion Beam Analysis.
ICM	Intramolecularly Cross-Linked Macromolecules. → Microgel.
Ideal Fluid	→ Inviscid Fluid.
IDR	→ Imbedded Disc Retraction Method.
IFLM	→ Inorganic Fullerene-Like Material.
IFR	→ Imbedded Fiber Retraction Method.
IHP	Inner Helmholtz Plane, → Helmholtz Double Layer.
Iijima, Sumio (1939–)	A Japanese physicist and developer of high-resolution electron microscopes. He is known for his discovery of the carbon nanotube while working in the United States in 1991. He also coined the term carbon nanotube. *See* References [135, 136].
ILS	→ Ionization-Loss Spectroscopy.
Imbedded Disc Retraction Method	(IDR) A method for determining the interfacial tension between two polymer melts. A disc of one polymer is imbedded in a second polymer. Temperature is then increased so that the polymers melt and the disc retracts into the shape of a sphere. The shape changes are monitored as a function of time, and the interfacial tension is calculated from analysis of the viscous and interfacial forces. → Imbedded Fiber Retraction Method (IFR).
Imbedded Fiber Retraction Method	(IFR) A method for determining the interfacial tension between two polymer melts. A short, uniform fiber of one polymer is imbedded in a second polymer. Temperature is then increased so that the polymers melt and the fiber retracts into the shape of a sphere. The shape changes are monitored as a function of time, and the interfacial tension is calculated from analysis of the viscous and interfacial forces. → Imbedded Disc Retraction Method (IDR).

Dictionary of Nanotechnology, Colloid and Interface Science. Laurier L. Schramm
Copyright © 2008 WILEY-VCH Verlag GmbH & Co. KGaA, Weinheim
ISBN: 978-3-527-32203-9

Imbibition	The displacement of a nonwetting phase by a wetting phase in a porous medium or a gel; the reverse of drainage.
Immersional Wetting	The process of wetting when a solid (or liquid) that is initially in contact with gas becomes completely covered by an immiscible liquid phase. → Wetting, → Spreading Wetting, → Adhesional Wetting.
Impactor	The separation of larger-sized, heavier particles from smaller-sized, lighter particles in a gas stream by having the former strike a plate positioned in the flow path. The lighter particles follow the gas flow around the plate, which is called a collector. For a given set of operating conditions, particles larger than a certain size will be impacted and caught on the collector. The particle size cut-off is not exact due to the effects of turbulence and the existence of velocity gradients. Particles smaller than the cut-off size will remain in the gas stream. A series of impactors, used to separate particles into several size fractions, is called a cascade impactor. Elutriation and cyclones operate on a similar principle.
Impingement Separator	→ Separator.
Incidental Nanoparticles	Nanoparticles that have been formed as a by-product of a natural or artificial process, as opposed to being formed by design. Examples include nanoparticles formed during welding, milling, grinding, or combustion.
Inclined-Plate Settling	→ Lamella Settling.
Increase of Volume Upon Foaming	In foaming, 100 times the ratio of gas volume to liquid volume in a foam. Also termed the "foaming power".
Indentation Hardness	→ Hardness.
India Ink	A black ink comprising a colloidal dispersion of lampblack or carbon black particles, as the pigment, stabilized by a natural gum, and dispersed in an aqueous adhesive solution. Also termed Indian ink, or Chinese ink. → Lampblack, → Carbon Black.
Indian Ink	→ India Ink.
Indifferent Electrolyte	An electrolyte whose ions have no significant effect on the electric potential of a surface or interface, as opposed to potential-determining ions that have a direct influence on surface charge. This distinction is most valid for low electrolyte concentrations. Example: for the AgI surface in water $NaNO_3$ would be an indifferent electrolyte, but both Ag^+ and I^- would be potential-determining ions.
Indirect Flotation	→ Reverse Flotation.

Induced Gas Flotation	$\rightarrow$ Froth Flotation.
Induction Forces	Debye forces. $\rightarrow$ van der Waals Forces.
Inelastic Scattering	$\rightarrow$ Light Scattering.
Infinite Clusters	$\rightarrow$ Percolation.
Infrared Reflection-Absorption Spectroscopy	(IRAS) A surface vibrational spectroscopic technique for studying adsorbed molecules on crystals. The absorption of infrared radiation due to the adsorbates is determined after reflection from a plane substrate surface. *See also* Table 7.
Inherent Viscosity	In solutions and colloidal dispersions, the natural logarithm of the relative viscosity, all divided by the solute or dispersed-phase concentration. $\eta_{Inh} = C^{-1} \ln(\eta/\eta_o)$. In the limit of vanishing concentration it reduces to the intrinsic viscosity. Also termed the "logarithmic viscosity number". *See* Table 4.
Initial Knockdown Capability	$\rightarrow$ Knockdown Capability.
Initial Spreading Coefficient	$\rightarrow$ Spreading Coefficient.
Ink Bottle Pore	Describes a shape of dead-end pore, in which a narrow throat is connected to a larger pore body, in a porous medium. $\rightarrow$ Porous Medium.
Inner Helmholtz Plane	(IHP) $\rightarrow$ Helmholtz Double Layer.
Inner Potential	(1) In the diffuse electric double layer extending outward from a charged interface, the electrical potential at the boundary between the Stern and the diffuse layer is termed the "inner electrical potential". Synonyms include the "Stern layer potential" or "Stern potential". $\rightarrow$ Electric Double Layer, $\rightarrow$ Zeta Potential. (2) The electric potential inside a phase, in this case termed the "Galvani potential". $\rightarrow$ Galvani Potential, $\rightarrow$ Outer Potential, $\rightarrow$ Jump Potential.
Inorganic Nanotube	A non-carbon nanotube. $\rightarrow$ Carbon Nanotube.
Inorganic Fullerene-Like Material	(IFLM) A non-carbon nanoparticle having a fullerene-like structure. $\rightarrow$ Fullerene.
Inorganic Nanotube	A non-carbon nanotube.
INS	$\rightarrow$ Ion-Neutralization Spectroscopy.
In Situ	In science and engineering, *in situ* generally refers to an aspect of a reaction or process taking place where it normally occurs, as opposed to moving it to some other place. Examples include studying a phenomenon where it occurs, in a reaction mixture, or in a process vessel. $\rightarrow$ *Ex Situ*.

Integral Capacitance of the Electric Double Layer	$\rightarrow$ Capacitance of the Electric Double Layer.
Integral Length Scale	$\rightarrow$ Microturbulence.
Intercalation	The formation of a layer of one material between layers of another. Certain compounds can expand clay crystals through intercalation. With salts that are intercalated as the total salt, the process is termed "intersalation". Clay minerals containing an intercalation layer are also termed pillar interlayered clay minerals, pillared inorganic layered compounds (PILC), or composite clay nano-structures.
Interface	The boundary between two immiscible phases, sometimes including a thin layer at the boundary within which the properties of one bulk phase change over to become the properties of the other bulk phase. An interfacial layer of finite specified thickness can be defined. When one of the phases is a gas, the term "surface" is frequently used.
Interface Emulsion	An emulsion occurring between oil and water phases in a process separation or treatment apparatus. Such emulsions can have a high solids content and are frequently very viscous. In this case the term "interface" is used in a macroscopic sense and refers to a bulk phase separating two other bulk phases of higher and lower density. Other terms: "cuff layer", "pad layer", or "rag layer emulsions".
Interfacial Film	A thin layer of material positioned between two immiscible phases, usually liquids, whose composition is different from either of the bulk phases.
Interfacial Layer	The layer at an interface that contains adsorbed species. Also termed the "surface layer". $\rightarrow$ Adsorption Space.
Interfacial Polymerization	A means to microencapsulate species. For example, a lipophilic drug in an oil phase can be microencapsulated by adding a hydrophobic monomer and then using water and a surfactant to prepare an oil-in-water emulsion. Next, a hydrophilic monomer is dissolved in the aqueous phase. The two different monomers interact at the oil/aqueous interface to create polymer films that form the micro-capsule walls. The inverse applies to the microencapsulation of a hydrophilic drug.
Interfacial Potential	$\rightarrow$ Surface Potential.
Interfacial Rheology	$\rightarrow$ Surface Viscosity.
Interfacial Rheometer	$\rightarrow$ Surface Viscometer.
Interfacial Tension	$\rightarrow$ Surface Tension.

Interfacial Tension Methods	*See* Table 15.
Interfacial Viscometer	→ Surface Viscometer.
Interfacial Viscosity	→ Surface Viscosity.
Interferometry	An experimental technique in which a beam of light is reflected from a film. Light reflected from the front and back surfaces of the film travels different distances and produces interference phenomena, a study of which allows calculation of the film thickness.
Intermediate Pore	An older term, now replaced by "mesopore".
Intermicellar Liquid	An older term for the continuous (external) phase in micellar dispersions. → Continuous Phase, → Micelle.
Internal Phase	→ Dispersed Phase.
Internal Surface	In porous media the surface contained in pores and throats that are in communication with the outside space. → Molecular Sieve Effect. Media having internal porosity also have internal surface area that can be available for sorption reactions. → Activated Carbon.
Intersalation	→ Intercalation.
Intralipid	→ Lipid Emulsions.
Intramolecularly Cross-Linked Macromolecules	(ICM) → Microgel.
Intrinsic Viscosity	The specific viscosity divided by the dispersed-phase concentration in the limits of both the dispersed-phase concentration approaching infinite dilution, and the shear rate approaching zero ($[\eta] = \lim_{c \to 0} \lim_{\gamma \to 0} \eta_{sp}/C$). Also termed "limiting viscosity number". *See* Table 4.
Inverse Micelle	A micelle that is formed in a nonaqueous medium, thus having the surfactants' hydrophilic groups oriented inward away from the surrounding medium.
Inversion	The process by which one type of emulsion is converted to another, as when an O/W emulsion is transformed into a W/O emulsion, and vice versa. Inversion can be accomplished by a wide variety of physical and chemical means.
Invert Emulsion	A water-in-oil emulsion. This term differs from the term "reverse emulsion", which is used in the petroleum field.
Inverting Surfactant	A surfactant that can be added to an emulsion to quickly invert it.
Invert-Oil Mud	An emulsion drilling fluid (mud) of the water-in-oil (W/O) type, that has a high water content. → Oil-Base Mud, → Oil Mud.

Inviscid Fluid	An ideal fluid that has no viscosity. Such a fluid cannot support any applied shear stress and flows without any dissipation of energy. Also referred to as an "ideal fluid", "Pascalian fluid", or a "nonviscous fluid".
Ion Beam Analysis	(IBA) A non-destructive method for the determination of surface composition in which high-energy ion beams (from a radioactive source or particle accelerator) are used to produce backscattered ions, and emitted X-rays and gamma rays, whose energies are measured. The energies determined are used to identify specific elements in the bombarded sample. Also termed Energetic Ion Analysis (EIA). Nuclear reaction analysis (NRA) and ion-scattering spectroscopy (ISS) are based on analyzing the backscattered ions. Particle induced gamma-ray emission analysis (PIGEA) is based on analyzing the gamma-rays. Particle induced X-ray emission analysis (PIXEA) is based on atomic fluorescence and involves analyzing the X-rays. Typical depth resolutions are of the order of 10–20 nm. → Ion-Scattering Spectroscopy; *see* Table 7.
Ion Exchange	A special kind of adsorption in which the adsorption of an ionic species is accompanied by the simultaneous desorption of an equivalent charge quantity of other ionic species. Ion exchange is commonly used for removing hardness and other metal ions in water treatment. The ion-exchange media can be arranged to provide a specific selectivity. → Sorbent-Motivated Sorption.
Ionic Strength	A measure of electrolyte concentration given by $I = {}^1\!/_2 \sum c_i z_i{}^2$, where c_i are the concentrations, in moles per litre, of the individual ions, i, and z_i are the ion charge numbers.
Ionist	A term applied to the founders of the discipline of physical chemistry (Ostwald, van't Hoff and Arrhenius) and their students, who opened up new areas of research with their theory of electrolytic dissociation. *See* Reference [137].
Ionization-Loss Spectroscopy	(ILS) A technique related to photoelectron spectroscopy in which the emission of secondary electrons is studied and used for the determination of surface composition. *See also* Table 7.
Ion-Neutralization Spectroscopy	(INS) A surface technique in which low-energy inert gas ions are made to strike a surface and become neutralized by a charge-transfer process that leads to the ejection of electrons, which are detected. Information about both the surface and the adsorbed material can be gained. *See also* Table 7.
Ionomer	A polymer molecule that contains pendant ionic groups, usually at a level of 10 to 15 mass %. The polymer backbone can be hydrocarbon or fluorocarbon. Used in plastics, and membranes and as surface coatings.

Ion-Scattering Spectroscopy	(ISS) A scattering technique used for the determination of surface composition by scanning the surface with a monoenergetic ion beam. The energy of the scattered ions is related to the mass of the scattering atoms at the surface so that the masses of the surface atoms can be determined. The techniques employing low-energy ions (less than 10 keV) are termed low-energy ion-scattering spectroscopy (LEIS), or ion-scattering spectroscopy (ISS). The techniques employing high-energy ions (greater than 100 keV) are termed medium-energy ion-scattering (MEIS) or high-energy ion-scattering (HEIS) spectroscopy. HEIS is also known as Rutherford Backscattering Spectroscopy (RBS). → Ion Beam Analysis; *see* Table 7. *See* Reference [57] for specific terms in ISS.
Ion-Selective Membrane	A membrane that is permeability-selective (permselective) for certain ions. Typically such a membrane will carry an electric charge and therefore tend not to be permeable to ions of like charge. Selectivity among ions of opposite charge to the membrane but like charge among each other can sometimes be achieved through adjustment of a membrane's pore sizes.
IRAS	→ Infrared Reflection-Absorption Spectroscopy.
Iridescent Layers	→ Schiller Layers.
Isaphroic Lines	Contours of equal foam stability plotted on foam-phase diagrams. Example: *See* Reference [8], p. 312.
Isobar	The mathematical representation of a phenomenon occurring at constant pressure. → Adsorption Isotherm.
Isodisperse	→ Monodisperse.
Isoelectric	An ionic macromolecule that exhibits no electrophoretic or electro-osmotic motion.
Isoelectric Focussing	A method for the separation of charged colloidal particles or large molecules. An electric field gradient is imposed along a supporting medium as in zone electrophoresis. In this case, however, the supporting medium also supports a pH gradient. A sample of mixture to be separated is applied to one end of the supporting medium, and electrophoretic motion of each species occurs until it comes to rest at a pH corresponding to its isoelectric point. Regions of different components separate along the direction of the electric field and pH gradient according to the different isoelectric points of the components (typically the cathode end is held at the most basic pH). → Zone Electrophoresis.
Isoelectric Point	The solution pH or condition for which the electrokinetic or zeta potential is zero. Under this condition a colloidal system will exhibit no electrophoretic or electro-osmotic motions. The

isoelectric point may or may not be identical to the isoionic point depending on the system concerned. → Isoionic Point, → Point of Zero Charge.

Isoionic	An ionic macromolecule system is isoionic if the only other ions in the system are the ions of the solvent, such as H^+ and OH^- in water.
Isoionic Point	The solution pH or condition for which a species has a zero net charge. Under this condition the species may truly have zero charge, but alternatively may have regions of opposite charges that balance and therefore zero net charge. Example: Proteins and kaolinite clay particles can exhibit isoionic points. → Isoelectric Point.
Isokinetic Sampling	Collecting samples of a flowing dispersion using a method in which the sampling velocity (in the sampling probe) is equal to the upstream local velocity. If these velocities are not the same (anisokinetic sampling) then fluid streamlines ahead of the probe will be distorted; collection of particles or droplets will be influenced by their inertia, which varies with particle size; and sampling will not be representative.
Isometric Particle	A particle that yields the same measurement in all three dimensions.
Isomorphic	Isomorphic sets are essentially identical in structure. In mineralogy, isomorphic substitution refers to the substitution of one cation for another in the mineral structure, when the mineral is forming. Example: isomorphic substitutions of Al for Si in the tetrahedral sheet and Mg for Al in the octahedral sheet are important in the formation of the clay mineral montmorillonite and represent the major source of permanent negative charge in this mineral's structure.
Isostere	The mathematical representation of a phenomenon occurring at constant volume. → Adsorption Isostere, → Adsorption Isotherm.
Isotachophoresis	A kind of capillary electrophoresis involving selected leading and trailing ions (cationic or anionic) chosen such that the mobilities of solutes of interest are intermediate with respect to those of the leading and trailing ions. Solutes become concentrated in a zone behind the leading ion front. Also termed "displacement electrophoresis". Used to analyze charged molecules in complex mixtures. *See* Reference [138].
Isotactic Polymer	→ Atactic Polymer, → Tacticity.
Isotherm	The mathematical representation of a phenomenon occurring at constant temperature. → Adsorption Isotherm.

Israelachvili, Jacob (Nissim)   An Israeli-born American chemical engineer and colloid scientist
(1944–)   known for his work on intermolecular and intersurface forces. He
                                 developed the Surface Forces Apparatus for directly measuring the
                                 forces between surfaces in fluids. He is the author of the classic
                                 textbook "Intermolecular and Surface Forces" (1991).

ISS   → Ion-Scattering Spectroscopy.

J

Jamin Effect	The ability of a capillary tube, or a channel in porous media, containing alternating liquid slugs and gas bubbles, to sustain a significant pressure without movement of the gas bubbles. It is by this mechanism that the introduction of gas bubbles into small capillary vessels can block the circulation of blood, or that the introduction of a foam into porous media can block further fluid flow. Also termed water lock effect.
Jar Test	For emulsions, foams, or water treatment, $\rightarrow$ Bottle Test.
Jet Impingement	A dispersion technique in which a jet of liquid is directed at a surface or at a jet of another liquid.
Jet Mill	A machine for the comminution, or size reduction, of mineral or other particles. Such machines accelerate feed particles in a jet and cause size reduction by promoting interparticle and particle-wall collisions at high speed. Very small-sized particles can be produced with these mills. Also termed jet pulverizers.
Jet Pulverizer	$\rightarrow$ Jet Mill.
Jones-Ray Effect	The decrease in apparent surface tension of water, as determined by capillary rise, due to the addition of small amounts of electrolyte. In reality the surface tension of water increases.
Joule, James Prescott (1818–1889)	A British natural philosopher who made contributions in the areas of heat and thermodynamics. His work led to the principle of conservation of energy. Other epnoyms include the "Joule-Thomson effect" (gas expansion and temperature), the "Joule effect" (heat produced in a wire by electric current), and the unit of "Joule" (for energy).

Dictionary of Nanotechnology, Colloid and Interface Science. Laurier L. Schramm
Copyright © 2008 WILEY-VCH Verlag GmbH & Co. KGaA, Weinheim
ISBN: 978-3-527-32203-9

Jump Potential

The difference between the inner (Galvani) potential and the outer (Volta) potential. That is, $1/e$ multiplied by the work required to bring unit charge from just outside a phase into that phase. Also termed "surface potential jump" or "chi potential". $\rightarrow$ Inner Potential, $\rightarrow$ Outer Potential.

K

Kataphoresis	$\rightarrow$ Electrophoresis.
Keesom Forces	$\rightarrow$ van der Waals Forces.
Keesom, Willem Hendrik (1876–1956)	A Dutch physicist best known to colloid and interface scientists for his work in molecular interactions, including the dipole-dipole interactions that are named for him. Keesom interactions form one of the kinds of van der Waals forces between any two bodies of finite mass. Keesom is also known for his work in cryogenics, as part of which he was the first to be able to solidify helium.
Kelvin Equation	An expression for the vapor pressure of a liquid droplet, $RT \ln (p/p_o) = 2\gamma V/r$, where ρ is the vapor pressure of the liquid in bulk, p_o is the vapor pressure of the droplet, γ is the surface tension, V is the molar volume, and r is the radius of the liquid droplet. $\rightarrow$ Young-Laplace Equation. The Kelvin equation predicts that small liquid droplets will evaporate at a greater rate than will large droplets and that small particles will be more soluble than larger particles.
Kelvin, Lord	$\rightarrow$ Thomson, William.
Kelvin Tetrakaidecahedron	A model polyhedron of the type that may represent that existing in foams and froths. The Kelvin tetrakaidecahedron has eight non-planar hexagon faces and six planar quadrilateral faces, and has been used to represent possibly one of the more stable foam cell shapes, in terms of minimizing surface free energy.
Keratin	Insoluble protein from hair, nails, and outer layers of skin. $\rightarrow$ Collagen.
Kinematic Eddy Viscosity	$\rightarrow$ Eddy Kinematic Viscosity.
Kinematic Viscosity	Kinematic viscosity is the absolute viscosity of a fluid divided by the density. *See* Table 4.

Dictionary of Nanotechnology, Colloid and Interface Science. Laurier L. Schramm
Copyright © 2008 WILEY-VCH Verlag GmbH & Co. KGaA, Weinheim
ISBN: 978-3-527-32203-9

Kinetic Coefficient of Friction	$\rightarrow$ Friction.
Kinetic Stability	Although most colloidal systems are metastable or unstable with respect to the separate bulk phases, they can have an appreciable kinetic stability. That is, the state of dispersion can exist for an appreciable length of time. Colloidal species can come together in very different ways; therefore, kinetic stability can have different meanings. A colloidal dispersion can be kinetically stable with respect to coalescence but unstable with respect to aggregation. Or, a system could be kinetically stable with respect to aggregation but unstable with respect to sedimentation. It is crucial that stability be understood in terms of a clearly defined process. $\rightarrow$ Colloid Stability, $\rightarrow$ Thermodynamic Stability.
Klevens Constants	The empirical parameters in an equation advanced by Klevens for predicting the critical micelle concentrations (cmc) of surfactants in terms of the number of carbon atoms in the hydrocarbon chain (n): $\log$ (cmc) $= A - Bn$, where A and B are the Klevens constants. The Klevens constants for numerous surfactants are tabulated in References [22, 23].
Kn	$\rightarrow$ Knudsen Number.
Knockdown Capability	A measure of the effectiveness of a defoamer. First, a column of foam is generated in a foam stability apparatus and the foam height is recorded. A measured amount of defoamer is added, and the reduction in foam height over a specified time period, for example, 2 s, is noted. The knockdown capability is the reduction in foam height. This test has many variations. Sometimes referred to as "initial knockdown capability".
Knockout Drops	Demulsifier that can be used to enhance the separation of oil from water and solids, in an emulsion, in the centrifuge test for determining basic sediment and water (BS&W). Also termed "slugging compound".
Knudsen Number	(Kn) The ratio of the mean free path for gas molecules to droplet or particle radius, for an aerosol of liquid droplets or solid particles. For aerosols having Kn $\gg$ 1 such properties as momentum, energy and mass transfer can be approximated by considering the particles or droplets to be essentially large gas molecules. $\rightarrow$ High-Dispersed Aerosols, $\rightarrow$ Aerosol Knudsen Number.
Köhler Illumination	In microscopy, the illumination provided through a condenser lens system, which is adjusted to produce optimum brightness with uniform illumination of a sample.
Kolmogorov Turbulence Microscale	$\rightarrow$ Microturbulence.

Krafft Point	(1) The temperature (in practice a narrow range of temperatures) above which the solubility of a surfactant increases sharply (micelles begin to be formed). Below the Krafft point only single, unassociated surfactant molecules (monomers) or ions (ionomers) can be present, up to a given solubility limit. Above the Krafft point, a solution can contain micelles and thus allow much more surfactant to remain in solution in preference to precipitating. Numerous tabulations are given in References [22, 23]. (2) In the soap industry the Krafft point is sometimes defined as the temperature at which a transparent soap solution becomes cloudy upon cooling [41].
Krafft Temperature	$\rightarrow$ Krafft Point.
Krafft, (Wilhelm Ludwig) Friedrich	A German chemist known for his discovery that the solubility in water of soaps increases dramatically when temperature increases beyond a transition temperature known as the Krafft Point. This is now known to be due to transition into temperatures for which added surfactant associates to form micelles. Krafft also discovered that soap solutions clean by colloidal suspension of dirt particles. *See* References [139, 140].
Krefeld Fabric	Refers to standard test fabrics (Krefeld standard fabric) produced by the Wäschereiforschung Krefeld (WfK) Cleaning Technology Research Institute in Krefeld, Germany. Krefeld standard fabrics, together with standard dirt materials, are used to test detergents and cleaning machines and to study detergency and the kinetics of soil removal.
Krieger-Dougherty Equation	An empirical equation for estimating the viscosity of a concentrated dispersion of particles. *See* Reference [141] and *see* Table 11.
Kugelschaum	$\rightarrow$ Gas Emulsion.

L

Lab-on-a-Chip	→ Micrototal Analysis System.

Lacing

In the food industry, the adhesion of a sheet of foam to a glass surface is termed lacing. Example: the sheet of draining foam that adheres to the inner wall of a glass, above the level of bulk liquid, as beer is consumed.

Lacquer

A protective surface coating product comprising a solution of a cellulose derivative, such as nitrocellulose or cellulose acetate, in solvent. Once applied as a surface coating, the solvent evaporates leaving behind a solid surface film. Lacquers may be pigmented or clear. → Varnish.

Lamella

→ Foam.

Lamella Division

A mechanism for foam lamella generation in porous media. Typically, when a foam lamella reaches a branch point in a flow channel, the lamella can divide into two lamellae rather than simply follow one of the two available pathways. This action is termed "lamella division". → Snap-Off, → Lamella Leave-Behind.

Lamella Leave-Behind

A mechanism for foam lamella generation in porous media. When gas invades a liquid- saturated region of a porous medium, it can not displace all of the liquid, but rather it leaves behind liquid lamellae that will be oriented parallel to the direction of the flow. A foam generated entirely by the lamella leave-behind mechanism will be gas-continuous. → Snap-Off, → Lamella Division.

Lamella Number

A dimensionless parameter used to predict the likelihood that a combination of capillary suction in plateau borders and the influence of mechanical shear will cause an oil phase to become

Dictionary of Nanotechnology, Colloid and Interface Science. Laurier L. Schramm
Copyright © 2008 WILEY-VCH Verlag GmbH & Co. KGaA, Weinheim
ISBN: 978-3-527-32203-9

emulsified and imbibed into foam lamellae flowing in porous media. It predicts that this will happen when $L > 1$, where $L = \Delta P_C / \Delta P_R = (\gamma^\circ_F)\ r_o/(\gamma_{OF})\ r_p$. Here, ΔP_C and ΔP_R are the pressure differences between inside a plateau border and inside the laminar part of a lamella, and the pressure difference across the oil-aqueous interface, respectively. r_o and r_p are the radii of the oil surface with which the lamella comes into contact, and that of the lamella plateau border, respectively. γ°_F and γ_{OF} are the foaming solution surface tension and the foaming solution-oil interfacial tension, respectively. Simplified forms of this equation have also been used. *See* Reference [142].

Lamellar Foam

Although all foams contain lamellae, this term is sometimes used to distinguish a certain kind of foam in porous media. When the length scale of the confining space is comparable with or less than the length scale of the foam bubbles, the foam is termed "lamellar foam" to distinguish from the opposite case, termed "bulk foam". $\rightarrow$ Foam, $\rightarrow$ Foam Texture.

Lamella Settling

A process for phase separation based on density differences. A commercial lamella settler for suspensions or emulsions comprises a stack of parallel plates spaced apart from each other and inclined from the horizontal. The space between each set of plates forms a separate settling zone. The feed is pumped into these spaces, at a point near the longitudinal middle of the plates. The less dense phases rise to the underside of the upper plates and flow to the tops of those plates. Meanwhile, the more dense phases settle down to the upperside of the lower plates and flow to the bottoms of those plates. Product is collected at the top of the plate stack, and tailings are collected at the bottom of the plate stack. Such an inclined lamella-settling process is much more efficient than vertical gravity separation. Also termed "inclined plate settling" or "inclined tube settling".

Lamina

$\rightarrow$ Foam.

Laminar Colloid

$\rightarrow$ Colloidal.

Laminar Flow

A condition of flow in which all elements of a fluid passing a certain point follow the same path, or streamline; there is no turbulence. Also referred to as "streamline flow".

Lampblack

Amorphous (gray-black) carbon particles created by the partial combustion of oil, coal tar, or other hydrocarbons. Lampblack is used as a pigment in paints, laser printer toner, crayons, carbon paper, matches, lubricants, rubber, and lead pencils. Also termed soot. $\rightarrow$ Carbon Black.

Landau, Lev (Davidovich) (1908–1968)

A Soviet physicist who worked with B.V. Derjaguin to develop a theory of the stability of colloidal particles that could explain the Schulze-Hardy rule. Independently developed also by Verwey and Overbeek the theory became famous as the "DLVO theory", an acronym constructed from the last names of these four scientists. Landau became more famous for his work in low-temperature and solid-state physics, and won the Nobel prize in physics in 1962 for his pioneering theories for condensed matter, especially liquid helium. Other eponyms include "Landau damping" (plasma physics) and the "Ginzburg-Landau theory" (low temperature physics). *See* References [143, 144].

Langmuir Adsorption

$\rightarrow$ Adsorption Isotherm.

Langmuir-Blodgett Film

A film of molecules that is deposited onto a solid surface by repeatedly passing the solid through a monolayer of molecules at a gas-liquid interface. Each pass deposits an additional monolayer on the solid. This process of creating a film on a surface is termed the Langmuir-Blodgett method. Example: The first applications of Langmuir-Blodgett films were in the preparation of nonglare coatings on glasses and other lenses.

Langmuir-Blodgett Method

$\rightarrow$ Langmuir-Blodgett Film.

Langmuir, Irving (1881–1957)

An American chemist and a founder of surface chemistry, Langmuir is best known to colloid and interface scientists for his contributions in the areas of adsorption (in which he identified chemisorption), monomolecular films and surface activity. He also made numerous other contributions in areas such as gaseous reaction kinetics and atmospheric and plasma physics (he coined the term "plasma"). Eponyms include the "Langmuir adsorption isotherm" and "Langmuir" (a scientific journal). He was awarded the Nobel prize (1932) in chemistry for his work in surface chemistry (the first industrial scientist to win the Nobel prize). *See* Reference [145].

Langmuir Isotherm

An adsorption isotherm equation that assumes monolayer adsorption and constant enthalpy of adsorption. The amount adsorbed per mass of adsorbent is proportional to equilibrium solute concentration at low concentrations and exhibits a plateau or limiting adsorption at high concentrations. $\rightarrow$ Adsorption Isotherm.

Langmuir Trough

$\rightarrow$ Film Balance.

Laplace, Marquis Pierre Simon de (1749–1827)

A famous French mathematician who made substantial contributions to astronomy and physics, he is familiar to colloid and interface scientists for his work on the theory of surface tension and the Young-Laplace equation. Other eponyms include "Laplace transforms" (mathematics) and the "Laplace ice calorimeter" (specific heat). *See* References [146, 147].

Laplace Flow	$\rightarrow$ Capillary Flow.
Laplace Waves	$\rightarrow$ Capillary Ripples.
Large Nuclei	Aerosol particles having diameters in the range 0.2 to 2.0 micrometres. Example: small pollens. *See also* Table 6, $\rightarrow$ Aerosol of Solid Particles.
Laser Ablation Processing	A fine particle synthesis method in which laser energy is used to erode material from a feedstock surface.
Laser Pyrolysis Gas-Phase Synthesis Method	A gas-phase fine particle preparation method in which a laser is used to rapidly heat a flowing reactive gas. The reaction product goes through nucleation and then growth through condensation, coagulation, and/or coalescence.
Lateral Force Microscopy	$\rightarrow$ Friction Force Microscopy.
Latex	A dispersion (suspension or emulsion) of polymer in water. Latex rubber, a heavily cross-linked polymer solid, is produced either by coagulating natural latex or by synthetic means through emulsion polymerization. Example: Latex paint is a latex containing pigments and filling additives. $\rightarrow$ Rubber Latex.
Lather	A foam produced by mechanical agitation on a solid surface. Example: the mechanical generation of shaving foam (lather) on a wet bar of soap.
Launderometer	The specialized machine used to perform a standardized test method for measuring the effectiveness of detergents. The degree to which reference soils are washed from standard fabric swatches in the presence of detergents and under specified conditions is determined. $\rightarrow$ Detergent, $\rightarrow$ Detergency.
Lava	$\rightarrow$ Magma.
Leave-Behind	$\rightarrow$ Lamella Leave-Behind.
LEED	$\rightarrow$ Low-Energy Electron Diffraction.
LEIS	Low-Energy Ion-Scattering Spectroscopy. $\rightarrow$ Ion-Scattering Spectroscopy.
Lemlich Equation	An equation for predicting the conductivities of foams. *See* Table 10.
Lennard-Jones 6-12 Potential	$\rightarrow$ Lennard-Jones Potential.
Lennard-Jones Potential	A measure of the potential energy of interaction between two atoms. Also termed the "Lennard-Jones 6–12 potential".
Lens	(1) Physics: Any piece of material or device that concentrates or disperses an incident beam of light, sound, electrons, or other

radiation. Example: the curved pieces of glass that magnify an image formed in a microscope.

(2) Colloid: A nonspreading droplet of liquid at an interface is said to form a lens. The lens is thick enough for its shape to be significantly influenced by gravitational forces.

(3) Geology: A specific geological layer resembling a convex lens. Example: clay mineral lens.

Levitator

An instrument for dielectrophoresis studies in which the motion of dipolar colloidal species is observed in a vertically imposed, nonhomogeneous electric field. Normally, depending on the sign of the Clausius-Mossotti factor, species tend to move toward either the most intense or the least intense region of the electric field. In this case, adjusting the frequency of the oscillating electric field (changing the Clausius-Mossotti factor) achieves a state of levitation for which the species move neither up nor down. → Dielectrophoresis, → Clausius-Mossotti Factor.

Lifshitz–van der Waals Forces

→ van der Waals Forces.

Light Crude Oil

A naturally occurring hydrocarbon having a viscosity less than 10,000 mPa·s at ambient deposit temperature, and a density less than 934 kg/m^3 at 15.6 °C. *See* References [50–52].

Light Nonaqueous-Phase Liquid

(LNAPL) → Nonaqueous-Phase Liquid.

Light Scattering

Light will be scattered (deflected) by local variations in refractive index caused by the presence of dispersed species depending upon their size [4]. In elastic scattering wavelength does not shift, in inelastic scattering wavelength shifts are caused by molecular transitions, and in quasielastic scattering wavelength shifts and line broadening are due to time-dependent processes. In light scattering the scattering plane contains both the incident light beam and the line that connects the center of the scattering system to the point of observation. The scattering angle lies in this plane and is measured clockwise by viewing into the incident beam. By this measure forward scattering is at a scattering angle of zero and backscattering is at a scattering angle of 180°. Sometimes termed "conventional light scattering" (CLS). → Dynamic Light Scattering, → Mie Scattering, → Nephelometry, → Rayleigh Scattering, → Tyndall Scattering.

Limiting Capillary Pressure

For foam flow in porous media the maximum capillary pressure that can be attained by simply increasing the fraction of gas flow. Foams flowing at steady-state do so at or near this limiting capillary pressure. In the limiting capillary pressure regime the steady-state saturations remain essentially constant.

Limiting Collision Efficiency	For perikinetic aggregation, the fraction of collisions that result in particle attachment, as given by the Smoluchowski equation. If there are particle-particle interactions the actual collision efficiency will be lower. → Aggregation Time.
Limiting Diffusion Coefficient	→ Diffusion Coefficient.
Limiting Sedimentation Coefficient	→ Sedimentation Coefficient.
Limiting Viscosity Number	→ Intrinsic Viscosity.
Line Tension	The reversible work per unit change of length of the periphery of a film (meniscus). The line tension is equivalent to a force tangential to the film, and is the one dimensional analogue of interfacial tension. For example, where three phases meet a line tension can exist along the three-phase junction. For a lens of material at the interface between two other immiscible phases, the three-phase contact junction takes the form of a circle along which the line tension acts. Also termed "Film Line Tension".
Lipid	Long-chain aliphatic hydrocarbons and derivatives originating in living cells, including fatty acids and closely related compounds, such as mono-, di-, and triglycerides (fats), and phospholipids. Some lipids, such as fatty acids, are also surfactants and self-assemble into micelles. Simple lipids tend to be hydrocarbon-soluble but not water-soluble. Examples include the acids: palmitic (from palm oil), oleic (cis-9-octadecenoic acid, from olive oil), linoleic and linolenic (from linseed oil), ricinoleic (from castor oil), and arachidic (from ground-nut oil). *See* References [9, 148]. *See also* Table 14.
Lipid Emulsions	Oil-in-water (O/W) emulsions that can be compatible with blood and therefore can be used for intravenous injection of nutritional fats, lipophilic drugs, and vitamins. Typical stabilizers for the fatty acid oil droplets are phospholipids. Lipid drug emulsions can deliver such agents as oxygen, sedatives, anesthetics, vitamins, platelet inhibitors, and diagnostic imaging agents. Example: Intralipid was probably the first lipid emulsion for clinical nutrition, and was designed to deliver energy and fatty acids to patients who cannot eat.
Lipid Bilayer	→ Bimolecular Film.
Lipid Film	A thin film of oil in water in which the film is stabilized by lipids. The term is used even though the film is not a film of lipid. → Fluid Film.
Lipophile	That part of a molecule that is organic-liquid-preferring in nature.

141 London, Fritz (1900–1954)

Lipophilic	The (usually fatty) organic-liquid-preferring nature of a species. Depending on the circumstances can also be a synonym for "oleophilic". → Hydrophile-Lipophile Balance.
Lipophobe	That part of a molecule that is organic-liquid-avoiding in nature.
Lipophobic	The (usually fatty) organic-liquid-avoiding nature of a species. Depending on the circumstances can also be a synonym for "oleophobic".
Liposome	→ Vesicle.
Liquid Aerosol	→ Aerosol.
Liquid Aphrons	→ Colloidal Liquid Aphrons.
Liquid-Crystalline Phase	→ Mesomorphic Phase.
Liquid Crystals	→ Mesomorphic Phase.
Liquid Crystal Templating	A means of preparing microporous or mesoporous inorganic oxides. The surfactant liquid crystals, chosen for their size and shape, form a template. Inorganic species, such as silicates, are deposited between the liquid crystals and then condensed to form a network. The liquid crystals are then removed, for example, by burning them off during calcination at elevated temperature, leaving the desired porous inorganic structure. *See* Reference [149].
Liquid Film	Four kinds of thin liquid films are typically distinguished, depending on the nature of the two phases bounding the film: (1) Foam films separating two vapour phases. (2) Emulsion films separating two droplets. (3) Suspension films separating two solid surfaces. (4) Wetting films separating a solid or liquid from a vapour.
Liquid Limit	The minimum water content for which a small sample of soil or similar material will barely flow in a standardized test method [43, 44]. Also termed the "upper plastic limit". → Atterberg Limits, → Plastic Limit, → Plasticity Number.
LNAPL	Light Nonaqueous-Phase Liquid. → Nonaqueous-Phase Liquid.
Logarithmic Viscosity Number	→ Inherent Viscosity.
London Forces	→ van der Waals Forces. → Dispersion Forces for a comment on the origin of this unfortunate synonym.
London, Fritz (1900–1954)	A German-born American philosopher, physicist and physical chemist, London is known to colloid and interface science for his work on intermolecular forces, and in particular for London forces: the $1/r^6$ induced dipole-induced dipole interaction (dispersion) forces.

Loose Emulsion	A petroleum industry term for a relatively unstable, easy-to-break emulsion, as opposed to a more stable, difficult-to-treat emulsion. → Tight Emulsion.
Lorenz-Mie Scattering	→ Mie Scattering.
Lotus Effect	Two natural examples of super-hydrophobicity include the surfaces of the leaves of the Lotus and Lady's Mantle plants. Water droplets bead-up completely on these leaves and remove dirt from them as they roll off. Accordingly, this self-cleaning effect is termed the Lotus effect. → Super-Hydrophobic Surface.
Low-Calorie Spread	→ Spreadable Fats.
Low-Fat Spread	→ Spreadable Fats.
Low-Energy Electron Diffraction	(LEED) A diffraction technique in which a low-energy electron beam is used. In this case the electrons have low penetrating power, and measuring the diffraction pattern yields information about surface structure. The inelastically scattered electrons, having lower energy, are stopped by grids, and the elastically scattered electrons, having the original energy level, are used to form the image pattern. Hence the term "elastic low-energy electron diffraction (ELEED)" is also used. → High-Energy Electron Diffraction, *see* Table 7.
Low-Energy Ion-Scattering Spectroscopy	(LEIS) → Ion-Scattering Spectroscopy.
Low-Energy Surface	Qualitative categorization indicating that a surface has a relatively low surface free energy (usually for van der Waals bonded materials). Example: paraffin wax. In contrast, high energy surfaces are those having relatively low surface free energies (usually for ionic or covalently bonded materials). → High-Energy Surface.
Lower-Phase Microemulsion	A microemulsion that has a high water content and is stable while in contact with a bulk oil phase, and in laboratory tube or bottle tests tends to be situated at the bottom of the tube, underneath the oil phase. For chlorinated organic liquids, which are denser than water, the oil is at the bottom phase rather than the top. → Microemulsion, → Winsor-Type Emulsions.
Lower Plastic Limit	→ Plastic Limit.
Lubrication	The action of a substance to reduce friction between two materials. Usually the lubricating film is thick enough for the material surfaces to be quite independent of each other. Boundary lubrication refers to the situation where only a thin film separates the material surfaces and the coefficient of friction depends upon the specific nature of the interfacial region. The science of friction and lubrication is known as "tribology".

Lundelius Rule	An expression for the inverse relation between solubility and the extent of adsorption of a species.
Lyocratic	A dispersion stabilized principally by solvation forces. Example: The stability of aqueous biocolloid systems can be explained in terms of hydration and steric stabilization. $\rightarrow$ Electrocratic.
Lyophilic	General term referring to the continuous-medium- (or solvent)-preferring nature of a species. $\rightarrow$ Hydrophilic.
Lyophilic Colloid	An older term used to refer to single-phase colloidal dispersions. Examples: polymer and micellar solutions. Other synonyms no longer in use: "semicolloid" or "half-colloid".
Lyophobic	General term referring to the continuous-medium- (or solvent)-avoiding nature of a species. $\rightarrow$ Hydrophobic.
Lyophobic Colloid	An older term used to refer to two-phase colloidal dispersions. Examples: suspensions, foams, emulsions.
Lyophobic Mesomorphic Phase	$\rightarrow$ Mesomorphic Phase.
Lyoschizophrenic Surfactant	A surfactant in a two-phase system whose behavior indicates a lack of preference for solubility in one phase or the other [15].
Lyotropic Liquid Crystals	$\rightarrow$ Mesomorphic Phase.
Lyotropic Mesomorphic Phase	$\rightarrow$ Mesomorphic Phase.
Lyotropic Series	A series and order of ions indicating, in decreasing order, their effectiveness in influencing the behavior of colloidal dispersions. Also termed "Hofmeister series". Example: The following series shows the effect of different species on coagulating power. Cations: $Cs^+ > Rb^+ > K^+ > Na^+ > Li^+$ Anions: $CNS^- > I^- > Br^- > Cl^- > F^- > NO_3^- > ClO_4^-$

M

MacMichael Viscometer	An early concentric cylinder-type rheometer in which the cup was rotated while the torque on a bob was measured. The torque was measured by reading from a scale divided into 300 units called MacMichael degrees.
MacMichael Degrees	→ MacMichael Viscometer.
Macroemulsion	→ Emulsion. The term "macroemulsion" is employed sometimes to identify emulsions having droplet sizes greater than some specified value and sometimes simply to distinguish an emulsion from the microemulsion or micellar emulsion types.
Macrohardness	→ Hardness.
Macroion	A charged colloidal species whose electric charge is attributable to the presence at the surface of ionic functionalities.
Macromolecule	A large molecule composed of many simple units bonded together. Macromolecules can be naturally occurring, such as humic substances, or synthetic, such as many polymer molecules.
Macropore	→ Pore.
Macroscopic Film	→ Film.
Magma	Magmas are subterranean hot, fluid precursors to lava. Magmas can contain hot liquids, gases, and solids in all proportions and combinations, so they can represent emulsions, foams, suspensions, or any combination of these dispersions at the same time. Each of these kinds of dispersions can also be found in lavas. Example: when pressure is released during the upward flow of obsidian lava a foam is formed. If the foamed lava cools without breaking the result is pumice (stone).

Dictionary of Nanotechnology, Colloid and Interface Science. Laurier L. Schramm
Copyright © 2008 WILEY-VCH Verlag GmbH & Co. KGaA, Weinheim
ISBN: 978-3-527-32203-9

Magnetic Force Microscopy | (MFM) A form of scanning probe microscopy (and a variation on atomic force microscopy) in which a magnetic tip (usually coated with gold to prevent electrical interaction) is used, and which is used to probe any magnetic field above a sample surface. The tip is mounted on a cantilever so that magnetic force is translated into deflection that can be measured. The resulting images provide information about topography as well as magnetic properties of a surface. *See* Reference [150].

Magnetic Resonance Imaging | (MRI imaging, nuclear magnetic resonance imaging) A technique for imaging and quantifying the distribution of phases in multi-phase systems, including dispersions in porous media. The technique employs a homogeneous, static, high magnetic field with a superimposed, time-dependent, linear-gradient magnetic field so that the total magnetic- field strength depends on position in the sample. Resonance is induced with radiofrequency energy. In the imaging system the position-dependent resonance frequencies and signal intensities allow the determination of the concentration, chemical environment, and position of any NMR-active nuclei in the sample.

Magnetophoretic Mobility | The mobility of a paramagnetic or ferromagnetic particle moving under the influence of an external magnetic field. The magneto-phoretic mobility equals the particle velocity, relative to the medium, divided by the magnetic-field gradient at the location of the particle [151]. This definition is analogous to the definition of electrophoretic mobility.

Magneto-Rheological Colloids | (MR Colloids) Dispersions whose viscosity can change (by orders of magnitude) when exposed to a sufficiently large external magnetic field. The magnetic field causes the previously randomly oriented particles to quickly align and form chains and structures, causing a fluid suspension to become solid. The transition is reversible; the viscosity returns to low levels when the external field is removed. Example: suspensions of iron particles in oil, which are also termed ferrofluids. → Smart Colloids.

Main Active | The primary surfactant in a detergent formulation. → Detergent.

Marangoni Effect | In surfactant-stabilized fluid films, any stretching in the film causes a local decrease in the interfacial concentration of adsorbed surfactant. This decrease causes the local interfacial tension to increase (Gibbs effect), which in turn acts in opposition to the original stretching force. With time the original interfacial concentration of surfactant is restored. The time-dependent restoring force is referred to as the "Marangoni effect" and is a mechanism for foam and emulsion stabilization. The combination of Gibbs and Marangoni effects is properly referred to as the "Gibbs-Marangoni effect", but is frequently referred to simply as the "Marangoni effect".

Marangoni Elasticity	$\rightarrow$ Film Elasticity, $\rightarrow$ Marangoni Effect.
Marangoni Flow	Liquid flow in response to a gradient in surface or interfacial tension. $\rightarrow$ Marangoni Effect.
Marangoni Number	A dimensionless quantity (symbol Ma) used to characterize the onset of Marangoni flow, or instability. The critical value above which Marangoni instability appears is about 50 to 100. The Marangoni number can be defined in terms of the gradient of interfacial tension with either solute concentration or temperature. An example is given in Reference [152].
Marangoni Surface Elasticity	$\rightarrow$ Film Elasticity, $\rightarrow$ Marangoni Effect.
Marangoni Waves	$\rightarrow$ Capillary Ripples.
Margarine	A water-in-oil emulsion in which the oil may be animal and/or vegetable oils. Traditional margarines contained about 20% (v/v) water and 80% (v/v) oil but lower oil-content formulations exist as well. The water droplets are separated, and stabilized, by fat globules and fat crystals. Synonyms from older terminology include oleo oil and oleomargarine. *See* Reference [153]. $\rightarrow$ Shortening, $\rightarrow$ Spreadable Fats.
Marine Colloids	Any colloids derived from marine sources. Examples include the hydrophilic colloids (hydrocolloids) derived from various seaweeds, such as algin, and colloids derived from marine animals, such as chitin. $\rightarrow$ Seaweed Colloids, $\rightarrow$ Chitin.
Marine Snow	Large aggregates formed in a marine environment from such microorganisms as bacteria and other compnents, linked together by polysaccharide tendrils [244]. These aggregates can be as large as several millimetres in diameter.
Mark, Herman (Francis) (1895–1992)	An Austrian-born American chemist known for his work in macromolecular, or polymer, chemistry and X-ray crystallography. He contributed to the structural determination of natural polymers, developed new polymers, and has been considered by some to be the founder of polymer science. Eponyms include the "Staudinger–Mark–Houwink equation" (polymer solution viscosity). *See* References [154].
Mark–Houwink Equation	$\rightarrow$ Staudinger–Mark–Houwink Equation, *see* Table 11.
Marsh Funnel Viscosity	A parameter intended to approximate fluid viscosity and measured by a specific kind of orifice viscometer. The measurement unit is Marsh seconds.

Martin's Diameter	A statistical particle diameter; the length of a line drawn parallel to a chosen direction such that it bisects the area of a particle. The value obtained depends on the particle orientation, and so these measurements have significance only when a large enough number of measurements are averaged together. $\rightarrow$ Feret's Diameter.
Mass-Area Mean Diameter	An average particle diameter calculated from measurement of average particle area.
Mass-Average Quantities	A method of averaging in which the sum of the amount of species multiplied by the property of interest squared is divided by the sum of the amount of species multiplied by the property. An example is the mass-average relative molecular mass determined by light-scattering methods,

$$M_{\mathrm{r,m}} = \frac{\sum n_i M_r^2(i)}{\sum n_i M_r(i)}$$

where n_i is the amount of species and $M_r(i)$ is the relative molecular mass of species i. $\rightarrow$ Number-Average Quantities.

Maximum Bubble Pressure Method	A method for the determination of surface tension in which bubbles of gas are formed and allowed to dislodge from a capillary tube immersed in a liquid. The maximum bubble pressure achieved during the growth cycle of the bubbles is used to calculate the surface tension on the basis of the pendant-drop analysis method. Variations include the differential maximum bubble pressure method, in which two capillaries are used and the difference in maximum bubble pressures is determined.
Maxwell-Wagner Polarization	A phenomenon in which the ions contained within a cell separate toward opposite sides of, but still within, the cell under the influence of an applied electric field. The charge separation within the cell creates a dipole.
MBS	$\rightarrow$ Molecular Beam Spectroscopy.
McBain, James William (1882–1953)	A Canadian physical chemist known for his contributions to colloid science, particularly in the area of surfactant (soap) solutions. His work led to the concept of the "association ion" or micelle. He introduced the phenonena and terminology of solubilization (of, for example, oils by incorporation into micelles) and cosolvency (solution by means of mixed liquid solvents). He also worked in the areas of oriented adsorption layers of surfactant, and the adsorption of liquids and vapours. He wrote textbooks on gas adsorption and colloid science and invented the McBain-Baker spring balance (for studying adsorption). *See* Reference [155].

μDMA	→ Differential Mobility Analysis.
MECC	→ Micellar Electrokinetic Capillary Chromatography.
Mechanical Alloying Method	An alloying process used to create fine particles through repetitive bonding, fracturing, and re-bonding of powdered materials in a mill, usually in the presence of an inert gas or a vacuum. This method can be used to create particles in the colloidal and/or nanoscale size ranges.
Mechanical Impact Mill	A machine for the comminution, or size reduction, of mineral or other particles. Such machines pulverize feed particles (typically about 10 mm initially) by causing them to strike a surface at high speed. Very small-sized particles can be produced with these mills.
Mechanical Syneresis	Any process in which syneresis is enhanced by mechanical means. → Syneresis.
Mechanosynthesis	→ Molecular Manufacturing, → Nanotechnology.
Medium-Energy Ion Scattering	(MEIS) → Ion-Scattering Spectroscopy.
Medium Sand	→ Sand, *see* Table 12.
Meerschaum	The German term for sea foam, Meerschaum is an older term for a solid foam: a dispersion of a gas in a solid. The term has also been used to refer to the mineral sepiolite when in rock, or bulk solid, form.
MEIS	Medium-energy ion-scattering spectroscopy. → Ion-Scattering Spectroscopy.
MELLFs	→ Metal Liquid-Like Films.
Membrane E.M.F.	→ Membrane Potential.
Membrane Filtration	→ Filtration.
Membrane Potential	The potential difference between two identical salt bridges placed into two ionic solutions that are separated from each other by a membrane. → Donnan Equilibrium.
Memory Devices	→ Molecular Circuitry.
MEMS	→ Microelectromechanical Systems.
Meniscus	The uppermost surface of a column of a liquid. The meniscus can be either convex or concave depending on the balance of gravitational and surface or interfacial tension forces acting on the liquid.
Mercury Porosimetry	→ Porosimeter.
Mesomorphic Phase	A phase consisting of anisometric molecules or particles that are aligned in one or two directions but randomly arranged in other directions. Such a phase is also commonly referred to as a

"liquid-crystalline phase" or simply a "liquid crystal". The meso-morphic phase is in the nematic state if the molecules are oriented in one direction; in the smectic state if oriented in two directions. Mesomorphic phases are also sometimes distinguished on the basis of whether their physical properties are determined mostly by interactions with surfactant and solvent (lyotropic liquid crystals) or by temperature (thermotropic liquid crystals). → Neat Soap.

Mesopore	→ Pore.
Mesoscopic Atoms	→ Quantum Dot.
Mesotechnology	→ Microtechnology, → Nanotechnology.

Metal Foam

A foam in which the continuous phase is a metal. Practical liquid metal foams contain colloidal-size particles and/or thin films at the liquid metal/gas interface. It is thought that these particles act to prevent or retard coalescence, possibly in similar fashion to the stabilization of aqueous foams by colloidal-sized particles. Example: solid aluminum foam prepared by cooling a liquid aluminum foam stabilized with alumina particles.

Metal Liquid-Like Films

(MELLFs) Surface films of coated silver nanoparticles that are highly reflective and behave like liquid mirrors. Nanometre-sized silver particles with adsorbed organic ligand, in a flocculated suspension, can be poured onto a substrate to create a liquid-mirror-like coating.

Metal Nanofoam	→ Nanofoam.
Metalworking Fluid	→ Grease.
Metastable	→ Thermodynamic Stability.
MFM	→ Magnetic Force Microscopy.
Micellar Aggregation Number	→ Aggregation Number.

Micellar Catalysis

Catalytic reactions conducted in a surfactant solution in which micelles play a role in catalyzing the reaction. Typically the micelles either solubilize needed reactant(s) or they provide a medium of intermediate polarity to enhance the rate of a reaction.

Micellar Charge

The net charge of surfactant ions in a micelle including any counterions bound to the micelle.

Micellar Electrokinetic Capillary Chromatography

(MECC) A kind of capillary electrophoresis involving micellar solutions and in which electroosmotic flow is maintained at a sufficient rate for all molecules to flow towards the cathode. Neutral and hydrophobic molecules partition between the micelles and the aqueous phase and are separated from each other. Example: used in amino acid analysis. *See* Reference [105].

Micellar Emulsion	An emulsion that forms spontaneously and has extremely small droplet sizes (<10 nm). Such emulsions are thermodynamically stable and are sometimes referred to as "microemulsions".
Micellar Mass	The mass of a micelle. For ionic surfactants, this value includes the surfactant ions and their counterions.
Micellar Solubilization	→ Solubilization.
Micellar Weight	→ Micellar Mass.
Micelle	An aggregate of surfactant molecules or ions in solution. Such aggregates form spontaneously at sufficiently high surfactant concentration, above the critical micelle concentration. The micelles typically contain tens to hundreds of molecules and are of colloidal dimensions. If more than one kind of surfactant forms the micelles, they are referred to as "mixed micelles". If a micelle becomes larger than usual as a result of either the incorporation of solubilized molecules or the formation of a mixed micelle, then the term "swollen micelle" is applied. → Critical Micelle Concentration, → Inverse Micelle.
Micelle-Mediated Extraction	→ Cloud Point Extraction.
Microcantilever	A microscale cantilever such as is used in any of the scanning probe microscopies. The cantilever beam may have nanoscale dimensions (nanobeam) and the term nanocantilever is sometimes used. → Scanning Probe Microscopy.
Micro-Differential Mobility Analysis	(μDMA) → Differential Mobility Analysis.
Microelectromechanical Systems	(MEMS) MEMS devices involve the integration of mechanical structures with microelectronics and are designed for specific purposes such as sensors and process controls. MEMS devices generally have at least one dimension in the 100 to 200 nm range. MEMS functions include microsensing, microactuating, microassaying, micromoving, and microdelivery. The science of MEMS includes the materials science aspects. MEMS examples include microfluidic chips. → Nanoelectromechanical System.
Microelectrophoresis	→ Electrophoresis.
Microemulsion	A special kind of stabilized emulsion in which the dispersed droplets are extremely small (<100 nm) and the emulsion is thermodynamically stable. These emulsions are transparent and can form spontaneously. In some usage a lower size limit of about 10 nm is implied in addition to the upper limit; → Micellar Emulsion. In some usage the term "microemulsion" is reserved for a Winsor type IV system (water, oil, and surfactants all in a single phase). → Winsor-Type Emulsions.

Microemulsion Polymerization	A polymer preparation method that involves free-radical polymerization in extremely small size, microemulsified monomer droplets. The produced polymer particles tend to be small and to have higher molar masses than are obtained from conventional emulsion polymerization. → Emulsion Polymerization, → Polymer Colloid.
Microencapsulation	The protection of a chemical species by containing it in small droplets, particles, or bubbles covered by a coating. Example: The encapsulation of liquid within vesicles.
Microfiltration	→ Filtration.
Microfluidic Chip	→ Biomedical Microelectromechanical Systems.
Microfluidics	The study of the behaviour of, and the control or manipulation of fluids on the microlitre scale. Microfluidics is used to manipulate such small volumes of fluids for such things as transport, separation, mixing, or measurement. The volumes concerned may be of the order microlitres, nanolitres (nanofluidics), or picolitres (picofluidics). Example: Micrototal analysis systems are a type of microfluidics devices. → Biomedical Microelectromechanical Systems, → Microelectromechanical Systems, → Micrototal Analysis System, → Nanoelectromechanical System.
Microfoam	→ Gas Emulsion.
Microgas Emulsions	A kind of foam in which the gas bubbles have an unusually thick stabilizing film and exist clustered together as opposed to either separated, nearly spherical bubbles or the more concentrated, system-filling polyhedral bubbles. The stabilizing film, sometimes called a "soapy shell," is thought to have inner and outer surfactant monolayers. A microgas emulsion will cream to form a separate phase from water. Also termed "aphrons" or "colloidal gas aphrons". → Colloidal Liquid Aphrons.
Microgel	Cross-linked clusters of polymer molecules that can usefully be considered to be distinct polymer molecules. Solutions of microgel molecules (or particles) exhibit different physical properties (like viscosities) than linear polymer coils. The term nanogel has been applied to relatively small, low molecular mass cross-linked clusters that have a radius of gyration in the nanoscale range (typically less than 10 nm) in solution. Both microgels and nanogels are also termed intramolecularly cross-linked macromolecules (ICM). *See* Reference [156].
Microhardness	→ Hardness.
Micrometre	10^{-6} m, a common distance unit in colloid science. The common symbol is "µm"; the symbol "µ" has also been frequently used, but is discouraged by the Systeme Internationale (SI). The milli-

micron, $10^{-3}\,\mu m$, has also been used, sometimes abbreviated as "mμ", also discouraged. The symbol "$\mu\mu$" has sometimes been used where millimicron, "mμ", was meant. $\rightarrow$ Micron.

Micrometre Thick Film $\rightarrow$ Film.

Micron   In the older literature, "micron" was one of three particle size range distinctions that were judged on the basis of visibility under the dark-field or bright-field microscope. Particles visible under the bright-field microscope were termed "microns" (diameters greater than about 500 nm). Particles not visible under the bright-field micro-scope but visible under the dark-field microscope (ultramicroscope) were termed "submicrons". Particles that were not even visible under the dark-field microscope were termed "amicrons" (diameters less than about 5–50 nm). These distinctions are no longer in use. The term "micron" to indicate 10^{-6} is discouraged by the Systeme Internationale; the correct term is "micrometre". $\rightarrow$ Micrometre.

Micronizing   The process by which a solid is reduced to particle sizes of less than about 100 μm by using any type of particle size reduction equip-ment. Examples: Micronized talc, micronized pigment.

Microphotograph   A photographic image forming a small copy of a much larger object. This image is not the same as a photomicrograph.

Micropore $\rightarrow$ Pore.

Micropore Filling   The process by which molecules become adsorbed within micropores.

Micropore Volume   The volume of adsorbed material that completely fills the micro-pores in a porous medium, expressed in terms of liquid volume at atmospheric pressure and specified temperature.

Microscopic Electrophoresis $\rightarrow$ Electrophoresis.

Microscopic Film $\rightarrow$ Film.

Microscopy   Light microscopy involves the use of light rays and lenses to observe magnified images of objects. The magnified image can be formed from transmitted light for transparent materials, or from reflected (incident) light for opaque materials. In each case there are different illuminating modes, and the light used can be visible, infrared, or ultraviolet. $\rightarrow$ Bright-Field Illumination, $\rightarrow$ Dark-Field Illumina-tion, $\rightarrow$ Köhler Illumination. Different viewing modes can be used, such as polarizing, fluorescence, phase contrast, and interference contrast. A derived technique is confocal microscopy. An analogous technique, electron microscopy, involves the use of electrons rather than light. $\rightarrow$ Confocal Microscopy, $\rightarrow$ Electron Microscopy.

Microsyneresis $\rightarrow$ Syneresis.

Microtechnology   A materials science term referring to materials or structures at a
                                   scale of about one micrometre (typically 1 to 100 µm), as opposed
                                   to the nanoscale (nanotechnology; 1 to 100 nm). The term meso-
                                   technology has been used to describe the regime in-between these
                                   two, although others are still using this term as a synonym for
                                   microtechnology. $\rightarrow$ Nanotechnology, $\rightarrow$ Molecular Nanotech-
                                   nology, $\rightarrow$ Picotechnology.

Microtome Method   A means of determining the surface concentration of species by
                                   cutting away a thin surface layer with a knife (microtome), physi-
                                   cally separating the layer, and analyzing it.

Micrototal Analysis System   (µTAS) A kind of microfluidics device comprising single- or
                                   multiple-chip analyzers that typically incorporate sample pretreat-
                                   ment, separation, and detection capabilities. These microma-
                                   chined devices were originally created using techniques derived
                                   from photolithography and applied to silicon substrates, but now
                                   additional fabrication methods such as injection moulding, and
                                   other substrates, such as glass and plastics are also used. Some of
                                   the benefits include integration of multiple techniques on a single
                                   chip, and miniaturization. In addition to microchannels, these
                                   systems may incorporate a number of microfluidic components:
                                   valves, pressure systems, metering systems, reaction chambers,
                                   and detection systems. Also termed "lab-on-a-chip," miniaturized
                                   analysis system, miniaturized total chemical analysis system,
                                   mTAS [157]. $\rightarrow$ Microelectromechanical Systems, $\rightarrow$ Micro-
                                   fluidics, $\rightarrow$ Nanoelectromechanical System.

Microturbulence   Refers to the turbulence associated with eddies that are so small
                                   they depend on the turbulent kinetic energy dissipation rate and on
                                   viscosity but are independent of the dimensions of the turbulence-
                                   generating system. There are three standard eddy size scales in
                                   turbulent flow. The integral length scale refers to the turbulence
                                   length scale for the "energy-containing eddies," those that contain
                                   most of the energy in the turbulence spectrum, and for which
                                   inertial dissipation is the predominant mechanism for turbulence
                                   kinetic energy dissipation. The Taylor turbulence microscale refers
                                   to the turbulence length scale (inertial subrange) for which viscous
                                   dissipation begins to affect the eddies. The Kolmogorov turbu-
                                   lence microscale refers to the turbulence (eddy) length scale for
                                   which energy dissipation by viscous forces is about the same as the
                                   dissipation by inertial forces. For larger eddies dissipation by
                                   viscous forces is unimportant. For smaller eddies dissipation by
                                   viscous forces is most important; this defines the turbulence
                                   microscale. The Reynolds number at the Kolmogorov microscale
                                   equals one.

Middle-Phase Microemulsion	A microemulsion that has high oil and water contents and is stable while in contact with either bulk oil or bulk water phases. This stability can be caused by a bicontinuous structure in which both oil and water phases are simultaneously continuous. In laboratory tube or bottle tests involving samples containing unemulsified oil and water, a middle-phase microemulsion tends to situate between the two phases. A bicontinuous microemulsion is sometimes termed a sponge phase. → Bicontinuous System, → Winsor-Type Emulsions.
Middle Soap	A mesomorphic (liquid-crystal) phase of soap micelles, oriented in a hexagonal array of cylinders. Middle soap contains a similar or lower proportion of soap (e.g., 50%) as opposed to water. Middle soap is in contrast to neat soap, which contains more soap than water and is also a mesomorphic phase, but has a lamellar structure rather than a hexagonal array of cylinders. Also termed "clotted soap". *See also* References [4, 41]. → Neat Soap.
Mie, Gustav (1868–1957)	A German physicist known to colloid and interface scientists for his contributions in the areas of light diffraction and colour effects. The scattering of light by species whose size is comparable with that of the incident light is termed "Mie scattering" (or "Lorenz-Mie scattering"). *See* Reference [158].
Mie Scattering	Light will be scattered (deflected) by local variations in refractive index caused by the presence of dispersed species, depending upon their size. The scattering of light by species whose size is much less than the wavelength of the incident light is referred to as "Rayleigh scattering", and it is termed "Mie scattering" if the species' size is comparable with that of the incident light. Also termed "Lorenz-Mie scattering". → Light Scattering, → Rayleigh Scattering.
Milk Solids-Not-Fat	(MSNF) The non-liquid and non-fat components in, or derived from, milk. These comprise lactose, casein micelles, whey proteins, minerals, vitamins, acids, and enzymes.
Milled Particle	A particle that has been produced by comminution (size reduction) of a solid in a mill.
Millimetre Thick Film	→ Film.
Millimicron	→ Micron.
Mineral Flotation	→ Froth Flotation.
Miniaturized Analysis System	→ Micrototal Analysis System.
Miniaturized Total Chemical Analysis System	→ Micrototal Analysis System.

Miniemulsion	→ Emulsion. The term is sometimes used to distinguish an emulsion from the microemulsion or micellar emulsion types. Thus a miniemulsion would contain droplet sizes greater than 100 nm and less than 1000 nm, or some other specified upper-size limit.
Mist	A dispersion of a liquid in a gas (aerosol of liquid droplets) in which the droplets have diameters less than a specified size. Mists are usually formed by condensation, but sometimes by atomization. In industry, mists have droplet sizes of less than 10 μm, as opposed to sprays, in which the droplet sizes are greater. In the atmosphere mists are aerosols of liquid droplets having sizes similar to those given for clouds in Table 5. → Fog.
Mist Drilling Fluid	→ Air Drilling Fluid.
Mitchell Foam Quality	→ Foam Quality.
Mixed gels	Mixed gels occur when two different components in a system each form a separate continuous network. There can be a synergy in the amounts of the materials needed to form a gel, in which such amounts are less than would be the case if they were used alone. Example: mixed gels formed from hydrocolloids and proteins.
Mixed Micelles	→ Micelle.
Mobile Film	→ Fluid Film.
Mobility Diameter	The effective spherical diameter of a particle determined based on its mobility. → Equivalent Spherical Diameter.
Mobility Reduction Factor	(MRF) A dimensionless measure of the effectiveness of a foam at reducing gas mobility when flowing in porous media. In one definition, the mobility reduction factor is equal to the mobility (or pressure drop) measured for foam flowing through porous media divided by the mobility (or pressure drop) measured for surfactant-free solution and gas flowing at the same volumetric flow rates.
Modifier	In flotation, a reagent added to alter surface properties in order to promote or reduce attachment of bubbles. Modifiers include chemicals intended to alter solution pH or to adsorb at surfaces and directly change surface charge and/or surface wetting properties. Example: simple modifiers include sodium hydroxide, sodium carbonate, and sodium silicate. → Collector; → Froth Flotation; → Frothing Agent.
Modulus of Surface Elasticity	→ Film Elasticity.
Mohs Scale	→ Hardness.
Molecular Beam Spectroscopy	(MBS) A technique for studying the kinetics of reactions at surfaces, in which a molecular beam strikes a surface and the lag time before which the appearance of reaction products is determined.

Molecular Circuitry	Nanoscale electronic circuit components including: molecular diodes, memory devices, regulators, switches, transistors, and wires. An early molecular switch was made from a molecular wire having alternating phenylene and ethylene groups, which allowed electric current to flow through it in one rotational conformation but not in another conformation. An early molecular circuit, or nanoscale circuit, was made from p- and n-type carbon nanotubes attached through gold electrodes with a conductive silicon gate electrode. → Nanowire, → Single-Electron Device.
Molecular Device	Also termed nanodevice. → Nanoelectromechanical System and → Biomedical Nanoelectromechanical Systems.
Molecular Diodes	→ Molecular Circuitry.
Molecular Disassembler	→ Disassembler.
Molecular Electronics	Refers to the design and construction of electronic devices consisting of small groups of molecules. Examples: molecular photovoltaic devices; molecular circuitry.
Molecular Engineering	→ Nanotechnology.
Molecular Fabrication	→ Nanotechnology.
Molecular Machine	A nanoscale machine; also termed nanomachine. → Molecular Manufacturing.
Molecular Manufacturing	The use of machines to direct the precise, controlled assembly of structures up from the molecular scale that are well-organized, and with reproducible properties. This is a component of molecular nanotechnology. The term mechanosynthesis is sometimes used to refer to molecular manufacturing where mechanical means are used to guide chemical reactions aimed at producing molecular structures. → Nanofabrication, → Nanotechnology.
Molecular Nanotechnology	The organization of atoms and molecules from the nanoscale up, in order to produce structures and/or materials having precise, predictable properties. Synonyms include molecular manufacturing, molecular engineering, molecular fabrication, mechanosynthesis, and chemosynthesis. The nanoscale structures created are sometimes termed nanoscale architectures, or nanoarchitectures. → Nanotechnology.
Molecular-Scale Electronics	→ Nanoelectronics.
Molecular Self-Assembly	→ Self-Assembly.
Molecular Sieve Effect	In porous media, the amount of internal surface accessible to molecules can depend on the size of the molecules and can be different for various components in a mixture. The different extent

of internal surface experienced by different molecules is termed the "molecular sieve effect". → Internal Surface.

Molecular Switch	→ Molecular Circuitry.
Molecular Systems Engineering	→ Molecular Manufacturing, → Nanotechnology.
Molecular Transistor	→ Molecular Circuitry.
Molecular Wire	→ Molecular Circuitry, → Nanowire.

Monodisperse

A colloidal dispersion in which all the dispersed species (droplets, particles) have the same size. Otherwise, the system is hetero-disperse (paucidisperse or polydisperse).

Monolayer Adsorption

Adsorption in which a first or only layer of molecules becomes adsorbed at an interface. In monolayer adsorption, all of the adsorbed molecules will be in contact with the surface of the adsorbent. The adsorbed layer is termed a "monolayer" or "monomolecular film".

Monolayer Capacity

In chemisorption, the amount of adsorbate needed to satisfy all available adsorption sites. For physisorption, the amount of adsorbate needed to cover the surface of the adsorbent with a complete monolayer.

Monomolecular Film	→ Monolayer Adsorption.
Monomolecular Layer	→ Monolayer Adsorption.

Monopolar

A polar substance that has only one kind of polar properties, either electron-donor or electron-acceptor.

Mooney Equation

An empirical equation for estimating the viscosity of an emulsion. *See* Table 11.

Mordant

(1) Textiles. In the dyeing of textiles a mordant is a substance that is strongly adsorbed onto the fibres and onto which dye(s) adsorb strongly. Mordant dyes are dyes that require a mordant, such as most natural dyes. Example: if alizarin is adsorbed onto cotton it creates a yellow colour that is easily washed off with detergent and water, however, if the cotton is first treated with alumina (mordant alumina) then with alizarin a red colour is produced that is not easily removed by detergent washing. Other mordants include tannic acid (for cotton), and metal salts such as potassium dichromate (for wool). (2) Histology. In the staining of tissue or other materials for viewing under a microscope, a mordant is a metal salt, usually involving trivalent aluminum or iron cations, that is adsorbed onto the material to be examined and to which a dye bonds.

Mordant Alumina

→ Mordant.

Mordant Dye	→ Mordant.
Motionless Mixer	→ Static Mixer.
Mousse Emulsion	→ Chocolate Mousse Emulsion.
Moving Boundary Electrophoresis	An indirect electrophoresis technique for particles too small to be visible. This principle is used in the Tiselius apparatus. Here a colloidal dispersion is placed in the bottom of a U-tube, the upper arms of which are filled with a less dense liquid that provides the boundaries and makes the connections to the electrodes. Under an applied electric field the motions of the ascending and descending boundaries are measured.
MRF	→ Mobility Reduction Factor.
MR Colloid	Also termed MR Fluid. → Magneto-Rheological Colloids.
MRI Imaging	→ Magnetic Resonance Imaging.
MSNF	→ Milk Solids-Not-Fat.
mTAS	Miniaturized Total Chemical Analysis System. → Micrototal Analysis System.
µTAS	→ Micrototal Analysis System.
Multilayer Adsorption	Adsorption in which the adsorption space contains more than a single layer of molecules; therefore, not all adsorbed molecules will be in contact with the surface of the adsorbent. → Brunauer-Emmett-Teller Isotherm, → Monolayer Adsorption.
Multiple Emulsion	An emulsion in which the dispersed droplets themselves contain even more finely dispersed droplets of a separate phase. Thus, there can occur oil-dispersed-in-water-dispersed-in-oil (O/W/O) and water-dispersed-in-oil-dispersed-in-water (W/O/W) multiple emulsions. These emulsions are sometimes called "double emulsions," "emulsions of emulsions", "three-phase emulsions", "triple-phase emulsions", or even "triple emulsions" (note that depending on the useage double emulsion and triple emulsion may refer to the same emulsion). More complicated multiple emulsions such as O/W/O/W and W/O/W/O are also possible.
Multi-Wall Carbon Nanotube	(MWCNT) → Carbon Nanotube.
Multi-Walled Nanotubes	→ Nanotube.
Mutual Coagulation	→ Heterocoagulation. → Aggregation.
MWNT	Multi-Wall Nanotube. → Carbon Nanotube.
Myelin Cylinders	Long-chain polar compounds, above their solubility limit, can interact with surfactants to form mixed micelles that separate

(as a coacervate) in the form of cylinders. These are termed "myelin cylinders" or "myelinic figures". They are usually quite viscous and can be birefringent.

Myelinic Figures

→ Myelin Cylinders.

Mysels, Karol (Joseph)
(1914–1998)

A Polish-born American chemist who specialized in colloid and interface science, Mysels is best known for his contributions in the areas of micellar systems (particularly regarding the sizes, shapes, structures, charges, and reactions of micelles), soap and foam films (especially their formation and stability), and many other areas in physical chemistry (especially adsorption at interfaces [10]). Among his eight published books are the textbook "*Introduction to Colloid Chemistry*," the monograph "*Soap Films*," and his 1971 book on critical micelle concentration (cmc) values [65], the latter of which is still much cited. Contrary to folklore the term micelle is not an eponym, although the fact that this view existed for some time speaks to his standing in the field. He was recognized by several national and International awards and served on a number of editorial boards including *Colloids and Surfaces* and *J. Colloid Sci.* *See* References [159, 160].

N

Nanite	→ Nanorobot.
Nanoaerosol	An aerosol of solid particles or liquid droplets having dispersed particles or droplets in the nanoscale range. The particles in a nanoaerosol are also termed ultrafine aerosol particles.
Nano Approach	→ Nanotechnology.
Nanoarchitecture	→ Molecular Nanotechnology.
Nanoarray	An organized, 2-D arrangement or array of nanoscale species or objects. One method of producing nanostructure involves self-assembly of molecules into organized structures, or nanoarrays. This approach has been used to make magnetic, semiconducting, and superconducting materials [38]. → Nanocluster.
Nanoassembler	→ Assembler; → Disassembler.
Nanobalance	A device for determining the mass of nanoscale materials. Example: a nanobalance has been made by attaching a species to the end of a carbon nanotube. The attached mass shifts the resonance frequency of the nanotube permitting measurement of the mass of the attached species. *See* Reference [161].
Nanobeam	(1) A nanoscale rod used in a mechanical device such as the cantilever in an atomic force microscope. (2) This term has also been variously applied to any of a number of kinds of focused beams, including electron, X-ray, and ion beams; generally in the context of surface analysis techniques.
Nanobelt	→ Nanoribbon, → Nanowire.
Nanobiology	→ Nanobiotechnology.
Nanobiosensor	→ Nanosensor.

Dictionary of Nanotechnology, Colloid and Interface Science. Laurier L. Schramm
Copyright © 2008 WILEY-VCH Verlag GmbH & Co. KGaA, Weinheim
ISBN: 978-3-527-32203-9

Nanobiotechnology	Biotechnology and a significant portion of biology have always involved aspects of the nanoscale. Nanobiotechnology refers more to nanotechnology, and especially molecular nanotechnology, in the context of or applied to biological systems. Examples include the use of biomolecules in the manufacture of nanodevices, and nanoscale probes for biological systems. Also termed bionanotechnology, nanobiology. → Biomedical Microelectromechanical Systems (BioMEMS).
Nanobot	→ Nanorobot.
Nanobottle	Nanoscale hollow spheres or capsules having a hole in the wall. Polymer hollow capsules have been prepared having pores that can open when the structure swells, while silica hollow spheres have been prepared with holes produced by calcination (silica nanobottles) [38].
Nanobubble	A nanoscale bubble. Transient nanobubbles sometimes occur when colloidal bubbles shrink but are normally quite unstable. Also termed ultrafine bubble.
Nanocage	A nanostructure in the form of a cage. Example: nanocages have been developed based on fullerenes, inorganic clathrate systems, mesoporous carbon, and self-assembling surfactant molecules. Nanocages can have an internal environment that is chemically different from that of bulk solution (as is the case with micelles), and in which chemical reactions can be conducted. Nanocages can also be used for encapsulation (nanoenapsulation). Also termed nanoshell, nanoflask. → Microencapsulation.
Nanocantilever	→ Microcantilever.
Nanocapsule	A hollow, usually spherical, structure at the nanoscale. Nanocapsules are generally made using polymer self-assembly and may be empty or filled with a solvent. Synonyms include colloidosome and hollow polymer nanostructure. Some of the applications for nanocapsules include using them as encapsulation (nanoencapsulation) media and as nanoreaction vessels.
Nanoceramic	Ceramic materials made from nanosize particles sintered together. Such materials can exhibit extremely high surface areas and are of interest for applications such as catalysis. Some non-oxide nanoceramics have shown the potential to combine heat-resistance with damage-resistance.
Nanoceramic Coating	A nanocoating comprising a ceramic nanopowder. Such coatings have been applied to increase corrosion resistance or wear-resistance. → Nanocoating.

Nanochemistry	The term usually refers to the use of synthetic chemistry to nanoscale building blocks, which could be of varying composition, structure, size, shape, and physical properties. *See* Reference [162].
Nanochip	A nanoscale electronic integrated circuit or photonic device.
Nanochondrion	A hypothetical nanomachine, or synthetic mitochondrion, that could be placed inside a cell and participate in its metabolism and/or reproduction. → Nanorobot.
Nanocluster	(1) A nanoscale aggregate of atoms or molecules. The term nanocluster is sometimes used to mean nanoarray, as in nanocluster array; and such clusters may have shapes that are dimensionally (D) essentially 1-D (nanowire), 2-D (nanoarray), or 3-D. → Nanoarray, → Nanowire.
	(2) The term nanocluster is also frequently used in a less technical sense, in reference to groups or networks of researchers or research institutions having common interests and/or facilities related to nanotechnology.
Nanocoating	The coating onto a surface of a nano-scale thick surface layer of physically or chemically bound material that is usually only a few molecules thick. Example: coating wood fibres with electrically or optically active polymers during paper making. There are a wide variety of other nanocoating applications including: antifogging, antimicrobial treatment, colour coating, UV protection, and wear-resistance. → Nanoceramic Coating.
Nanocomposite	A composite material that contains at least one phase having at least one dimension in the nanoscale range. Such nanoscale components are incorporated in order to improve the properties of the material in some way. Polymer nanocomposites typically incorporate silicates such as clays. Other nanocomposites incorporate metals, oxides, or carbon for example, all at the nanoscale. → Nanohole, → Nanophase.
Nanocomputer	A conceptual computing device (→ Drexler, (Kim) Eric) that would carry the instructions for the operations of assemblers, disassemblers, and replicators.
Nanocore	The interior portion of a nanoparticle that has been encapsulated in, or coated by, a dissimilar nanomaterial.
Nanocrane	→ Assembler; → Disassembler.
Nanocrystal	A crystal having one or more dimensions in the nanoscale range. Also termed nanocrystallite.
Nanocrystallite	→ Nanocrystal.

Nanodevice	Also termed molecular device. → Nanoelectromechanical System and → Biomedical Nanoelectromechanical Systems.
Nano-Differential Mobility Analysis	→ Differential Mobility Analysis.
Nanodisassembler	→ Assembler; → Disassembler.
Nanodispersions	Colloidal dispersions. → Colloidal, → Nanotechnology.
Nanodot	Nanoscale metal particles or clusters, usually embedded in an insulating material. Also termed nanograin. Nanodots of 1–2 nm diameter have been created. Example: magnetic nanodots with switchable polarity are a possible class of new memory elements for use in computing. → Quantum Dot.
Nanodroplet	A droplet whose diameter falls between 1 and 100 nm. Also termed ultrafine droplet, especially when dealing with liquid aerosols.
Nanoelectrochemical Patterning	→ Scanning Probe Surface Patterning.
Nanoelectromechanical System	(NEMS) The nanoscale equivalent of microelectromechanical systems (MEMS), which are microscale machines. NEMS are typically high frequency, high sensitivity devices for use in telecommunications, optical switches, and sensors. Also termed nanomechanical systems [37]. → Microelectromechanical Systems.
Nanoelectronics	The realm of electronics dealing with fabrication, study, and/or application of nanosized electronic devices.
Nanoemulsion	An emulsion with dispersed droplet sizes in the nanoscale range. Due to such very small droplet sizes many nanoemulsions are transparent. Synonyms include ultrafine emulsion, submicrometre emulsion. A nanoemulsion may also be a miniemulsion. Nanoemulsions differ from microemulsions in that the former are not thermodynamically stable. → Emulsion, → Miniemulsion.
Nanoencapsulation	Nanoscale encapsulation. → Microencapsulation, → Nanocage.
Nanofabrication	The fabrication of devices or materials with nanoscale precision. Current nanofabrication technologies are typically based on the thin film patterning techniques developed in the microelectronics industry, such as nanolithography. Other work is aimed at developing "bottom-up" molecular assembly techniques by which structures could be made. Also termed nanomanufacturing. → Molecular Manufacturing, → Molecular Nanotechnology, → Nanorobot.
Nanofatigue	The loss in material strength under loading, at the nanoscale. Nanohardness and nanofatigue measurements both involve nanoscale indentation (nano-indentation) and nanoscale indentation forces. → Hardness.

Nanofibre	Nanoscale fibre. Sometimes defined as a nanomaterial having two dimensions at the nanoscale and an aspect ratio of more than 3 to 1. Types of nanofibres include nanowire, nanorod, nanoribbon, and nanowhisker. $\rightarrow$ Carbon Nanotube, $\rightarrow$ Nanowire, $\rightarrow$ Platelet-Nanofibre.
Nanofilter	A filter with nanoscale pores. Also described as comprising a nanomesh. $\rightarrow$ Nanotube.
Nanofiltration	Pressure-driven membrane separation capable of selectively removing nanoscale materials from a fluid carrier. Nanofiltration membranes can typically separate species from about 0.5 nm in size to just under 10 nm in size. This range overlaps the typical upper limit for reverse osmosis and the typical lower limit for ultrafiltration. Nanofiltration can be used to reject dissolved multivalent metal ions while passing monovalent metal ions. $\rightarrow$ Filtration, $\rightarrow$ Nanofilter.
Nano-Flash Memory Device	A flash memory device that uses one or more nanoparticles as charge storage element(s). Charge is either added or removed from the charge storage element depending on the logic level.
Nanoflask	$\rightarrow$ Nanocage.
Nanofluidics	The study of the behaviour of, and the control or manipulation of fluids on the nanolitre scale. $\rightarrow$ Microfluidics, $\rightarrow$ Biomedical Microelectromechanical Systems, $\rightarrow$ Microelectromechanical Systems, $\rightarrow$ Micrototal Analysis System, $\rightarrow$ Nanoelectromechanical System.
Nanofoam	A solid foam, including xerogels and aerogels, in which the pore spaces have nanoscale dimensions. Examples include carbon nanofoams, silica nanofoams, and metal nanofoams. Carbon nanofoams are typically produced by sol-gel polymerization followed by pyrolysis, and are characterized by low density, continuous porosity, and high surface area, electrical conductivity, and capacitance. Also termed carbon aerogel. Silica nanofoams are typically produced by a sol-gel process and are of interest for applications including thermal insulating.
Nanogap	Fabricated nanoscale gaps between metal electrodes in nanoelectronic devices. Such gaps can be created by electrochemical deposition of conducting material, followed by back-etching to create the gap. Some very small nanogaps ($\sim$10 nm width) exhibit resistance changes in response to high applied bias voltages, due to changes in the gap width between the metal electrodes. Such properties are being used to create nanogap junctions with adjustable properties. The term nanogap has a different meaning from that of nanopore, $\rightarrow$ Pore.

Nanogel	(1) Small cross-linked clusters of polymer molecules that can usefully be considered to be distinct polymer molecules. The term nanogel has been applied to relatively small, low molecular weight cross-linked clusters that have a radius of gyration in the nanoscale range (typically less than 10 nm) in solution. Larger clusters are termed microgels. Both nanogels and microgels are also termed intramolecularly cross-linked macromolecules (ICM). → Microgel.
	(2) Nanogel is sometimes used to refer to aerogels containing nanoscale pores, and it is also a commercial tradename for an aerogel product line. With this meaning nanogel is sometimes referred to as frozen smoke.
Nanografting	→ Scanning Probe Surface Patterning.
Nanohardness	→ Hardness.
Nanohole	Holes on the nanoscale. Example: 10–100 nm diameter holes are created in some kinds of membranes. Nanocomposites can be created by filling nanoholes in polymer membranes with metals.
Nanohorn	A nanoscale cone, or horn having a curved longitudinal axis.
Nanoid	→ Nanorobot.
Nanoimprinting	→ Nanoimprint Lithography.
Nanoimprint Lithography	(NIL) A form of contact lithography in which a pattern is transferred from a rigid mold to a liquid/melt resist through UV-induced cross-linking or thermal embossing. Patterns with features in the 1–10 nm range have been made in this manner. → Dip Pen Nanolithography and → Nanophotolithography.
Nano-Indentation	→ Hardness.
Nanoindenter	A nanoindenter is an atomic force microscope used as a means of determining Young's modulus, hardness and/or yield strength by pressing the cantilever tip, or a modification thereof, into a surface. The word nanoindenter is also used commercially as a trademark. → Hardness.
Nanolithography	Lithography conducted at the nanometre scale. → Dip Pen Nanolithography, → Nanoimprint Lithography, → Nanophotolithography, → Scanning Probe Surface Patterning.
Nanology	The study of nanoscience, nanotechnology, and their origins and evolution. *See* Reference [163].
Nanolubrication	→ Nanotribology.
Nanomachine	→ Molecular Machine.

Nanomagnetic Sponge	A highly porous gel containing magnetic nanoparticles cross-linked into the gel structure. Developed as a tool to aid in the cleaning of works of fine art, the porous nature of the nanomagnetic "sponge" allows the material to hold and dispense cleaning solution or microemulsion. The magnetic nature allows the use of an external magnetic field to ensure that the sponge either does not contact the artefact or contacts it only lightly. *See* Reference [164].
Nanomanipulator	The probe used in scanning probe microscopies can be used to push against a surface with variable force. Coupling of the probe control to a human interface and an imaging system has been used to create a nanomanipulator. $\rightarrow$ Nanotweezers, and $\rightarrow$ Scanning Probe Microscopy.
Nanomanufacturing	$\rightarrow$ Molecular Manufacturing, $\rightarrow$ Nanofabrication.
Nanomaterial	Any material having one or more nanoscale dimensions (1 to 100 nm). Use of this term is also usually taken to imply that the material has different properties than it would if not for the nanoscale dimenesions.
Nanomesh	$\rightarrow$ Nanofilter.
Nanometre Thick Film	$\rightarrow$ Film.
Nanomotor	A nanoscale motor able to convert electrical or chemical energy into force and/or motion. One kind of rotational bearing has been made by attaching a gold plate to the outer shell of a suspended multiwall carbon nanotube and using electrostatic force to rotate the outer shell relative to the inner core. Such bearings have been used to support and drive small gold rotors. A biochemical example is ATPase, which essentially comprises rotating nanoturbines embedded in lipid membranes, and which power living cells. Here, the turning of the nanoturbine's shaft, caused by a proton imbalance across the membrane, permits the ATPase to convert adenosine diphosphate (ADP) to adenosine triphosphate (ATP). This kind of nanomotor is about 12 nm in diameter and turns at about 1,000 rpm. *See* References [165, 166].
Nanonail	Nail-shaped, nanoscale fibres or whiskers having the characteristic "nail-head" overhang on one end. Example: the nanonail-studded surfaces created to produce super-hydrophobic or super-lyophobic materials. *See* Reference [167]).
Nano-Onion	A nanoparticle comprising concentric molecular shell layers. Also termed Nested Nanoparticle.
Nanoparticle	Any solid particle having at least one dimension on the nanoscale (1 to 100 nm). Nanoparticles are frequently composed of a few hundred to a few thousand atoms. The gold sols of classical colloid

science were comprised of dispersed gold nanoparticles. A modern example is provided by the photo-oxidizing nanoparticles being incorporated into "self-cleaning" clothing. Also termed ultrafine particle, especially when dealing with aerosol particles such as in fumes. → Colloidal.

Nanopen

→ Dip Pen Nanolithography.

Nanopharmaceutical

A pharmaceutical product made using nanotechnology. Examples: pharmaceutical preparations that include nanoemulsions, nano-encapsulation, or nanoparticles.

Nanophase

A phase within a material, the former having at least one dimension in the nanoscale. → Nanocomposite.

Nanophotolithography

Photolithography conducted at the nanoscale. Example: one form of nanophotolithography uses a combination of optical lithography and polymer self-assembly to produce nanoscale patterning dimensions [168]. → Dip Pen Nanolithography and → Nanoimprint Lithography.

Nanophotonics

The realm of optics dealing with fabrication, study, and/or application of optical phenomena at the nanoscale.

Nanopipette

A nanoscale pipette used to dispense small amounts of fluid. Typically, a nanotube is attached to the end of a larger pipette, such as a glass micropipette. The nanotube tips usually have internal diameters of the order of 50 nm, and the nanopipettes can typically deliver volumes as small as a few femtolitres (10^{-15} L).

Nanoplotter

→ Dip Pen Nanolithography.

Nanopore

A nanoscale hole in a membrane, or a nanoscale hole or channel in a solid or gel. → Nanoporous, → Pore.

Nanoporous

Materials containing holes or channels that have diameters in the range 1 to 100 nm. Bulk nanoporous materials are of interest as highly efficient adsorbents, for example. Nanoporous membranes are of interest as a means of controlling molecular transport. Commercially available nanoporous membranes contain channels having diameters in the range 10 to 200 nm. → Pore, → Porous Medium, *see* Table 18.

Nanopowder

Dry nanoparticles.

Nanoprobe

(1) This term usually refers to the tip of a scanning-probe microscope.
(2) The term is sometimes used to refer to a nanoscale machine that is sent "in" to a system as an imaging, sensing, or nanorobotic device. → Nanosensor.

Nanoreplicator	→ Replicator.
Nanoresist	→ Dip Pen Nanolithography.
Nanorheology	The rheology of nanoscopically confined fluids, such as in molecular thin films. Nanoviscosities have been determined by direct methods involving molecular thin films and tools such as the atomic force microscope. Indirect methods have also been used, such as monitoring the Brownian motions (velocities) of nanoscale particles. The rheology and tribology of nanoscopically confined molecules have also been studied through molecular dynamics (numerical) simulation. → Nanotribology.
Nanoribbon	A solid nanofibre having an approximately rectangular cross-section, and for which the ratio of longer to shorter dimensions is more than 2 to 1. Also termed nanobelt. → Nanofibre, → Nanorod, → Nanowire.
Nanorobot	A conceptual robot device (→ Drexler, (Kim) Eric), autonomous or semi-autonomous, containing nanoscale components. Early prototypes of such devices are typically at the microscale, although attention is focused on being able to build them at the nanoscale. Also termed nanobot, nanoid, nanite.
Nanorod	A straight, solid, nanofibre. → Nanofibre, → Nanoribbon, → Nanowire.
Nanorope	Nanofibres bundled in a twisted conformation.
Nanoscale	Having at least one dimension in the range 1 to 100 nm, where a nanometre (nm) = 10^{-9} m. The nanoscale range is sometimes operationally extended to 0.1 to 100 nm, placing it just beyond the picoscale range. Also termed Nanosize.
Nanoscale Architectures	→ Molecular Nanotechnology.
Nanoscale Indentation	→ Hardness.
Nanoscience	A sub-discipline of science that emerged in the late 1990's and which overlaps with the size range of colloid science. Nanoscience involves the study of nanoscale materials, processes, phenomena, and/or devices. Nanoscience includes materials and phenomena at the nanoscale (typically 1 to 100 nm), hence it includes such areas as carbon nanoscience (e.g., fullerenes), molecular-scale electronics, molecular self-assembly, quantum size effects, and crystal engineering. → Colloid, → Colloid Science, → Nanotechnology.
Nanoscope	This term is used in a number of corporate and trademarked product names, including several atomic force and scanning probe microscopes.

Nanoscopic Film	→ Film.
Nanosensor	Nanoscale devices capable of detecting and/or measuring the concentration of specific chemicals or classes of chemicals. Typical nanosensors are optical or electrochemical in nature. A mechanical example is a nanocantilever coated with sensor molecules, which will bend when the sensor molecules bind to specific molecules on a substrate surface. Optical nanosensors use either a chemical or biochemical species to achieve the desired specificity of detection, hence the terms Chemical Nanosensor and Nanobiosensor. An example of a medical nanosensor is a nanoscale probe capable of being inserted into cells, attaching itself to a particle, and producing a magnetic field enabling it to be tracked.
Nanosheets	Nanoscale sheets of material. Example: graphene nanosheets as thin as a few atoms have been prepared from pyrolytic graphite [169].
Nanoshell	→ Nanocage.
Nanosize	→ Nanoscale.
Nanosizing	→ Attrition.
Nanosome	A nanoscale liposome. → Vesicle.
Nanostructure	Any organized structure that contains chemically or physically distinguishable components, at least one of which has one or more dimensions in the range from 1 to 100 nm. Example: An aggregate of nanoparticles is a nanostructured material.
Nanostructured Initiator Mass Spectrometry	(NIMS) A mass spectrometry method that uses surface activator molecules to initiate the desorption and ionization of sample molecules (as opposed to SIMS, which uses high-energy ions). The initiator molecules are initially trapped in pores on the surface of a specially prepared substrate onto which is placed the sample to be analyzed. When a laser or ion beam is used to vapourize the initiator molecules they also lift some of the sample molecules into the gas phase where they can be analyzed. Also termed nanostructure-initiator mass spectrometry.
Nanostructured Materials	→ Nanostructure.
Nanostructure-Initiator Mass Spectrometry	→ Nanostructured Initiator Mass Spectrometry.
Nanosuspension	A nanoscale suspension. → Suspension.
Nanotechnology	A rapidly growing area of materials science involving the design, characterization, manipulation, incorporation, and/or production of materials and structures in the nanoscale range (1 to 100 nm, sometimes operationally extended to 0.1 to 100 nm), by any of a

variety of physical and chemical methods [32, 170]. An important distinction is that these applications exploit the properties, distinct from bulk or macroscopic systems, of the nanoscale components. There is a substantial overlap of scale between nanotechnology and colloid technology. Although some nanodispersions are simply colloidal dispersions under a new name, some aspects of nano-technology are genuinely new, and have unusual properties, such as carbon nanotubes and quantum dots. Some mechanical aspects of nanotechnology deal with colloidal dispersions, such as the use of colloidal ink dispersions in robocasting to build near-nanometre scale three-dimensional structures. Nanotechnology also encompasses the 'nano' approach, or molecular nanotechnology, by which is meant the precise, controlled assembly of structures up from the molecular scale that are well-organized, and with reproducible properties. $\rightarrow$ Colloidal, $\rightarrow$ Microtechnology, $\rightarrow$ Nano-science, $\rightarrow$ Picotechnology.

Nano Test Tube $\rightarrow$ Nanotube.

Nanotribology   The study of friction, lubrication, and wear at the nanoscale. The nanoscale study of lubrication films, for example, has been conducted using variations of atomic force microscopy, such as friction force microscopy. $\rightarrow$ Nanorheology, $\rightarrow$ Nano-Triboscope.

Nano-Triboscope   A scanning-probe, friction-force instrument designed to measure friction and wear down to nanometre (and nanoNewton) scales. Also termed Tribolever. $\rightarrow$ Nanotribology.

Nanotube   A nanostructure in the form of a tube. That is, a hollow nanofibre. These can be made from carbon (carbon nanotubes), or a variety of other materials (inorganic nanotubes). Nanotubes can have metallic or semiconducting properties and can be good conductors of heat. Example: carbon nanotubes of about 2 nm diameter have been prepared and filled with water molecules, in which case the water molecules are arranged in essentially one-dimension only [171]. Such nanotubes have been used as nano test tubes, to provide highly confining reaction vessels. Nanotubes can be single-walled or multi-walled, the latter comprising concentric cylinders. Nanotubes can also be used as nanoscale filters to remove, for example, bacteria and viruses from water [172, 173]. $\rightarrow$ Carbon Nanotube.

Nanoturbine $\rightarrow$ Nanomotor.

Nanotweezers   Tweezer-like devices used to grasp and measure and/or manipulate objects at the nanoscale. Early nanotweezers comprised carbon nanotubes attached to each of two electrodes. By applying a voltage gradient to the electrodes, the nanotubes can be made to bend towards each other (like the motion of chopsticks) activating the tweezers. → Nanomanipulator, → Optical Tweezers, and → Scanning Probe Microscopy.

Nanowhisker   Nanoscale tendrils, or wires, grown from and/or extending from a substrate surface. Freestanding nanowhiskers on metal surfaces are of interest in electronics and photonics. → Tin Whisker.

Nanowire   Nanoscale wire, that is, a conducting or semi-conducting nanofibre. Nanowires, and other shapes including nanobelts, nanoribbons and nanorods, are typically used in nanoscale electronic circuit elements [174]. Nanowires, nanofibres, and nanotubes also provide building blocks for nanostructures of various kinds, and substrates for the development of new catalysts. Example: silicon nanowires have been prepared and then coated, to produce nanowires or semiconducting nanowires having diameters of about 10 nm [175, 176]. Coaxial nanowires have been prepared using a doped silicon core coated with a pure silicon layer, which in turn is coated with another doped silicon layer. Such coaxial nanowires have been used to connect nanoelectronic devices.

Naphtha   A petroleum fraction that is operationally defined in terms of the distillation process by which it is separated. A given naphtha is thus defined by a specific range of boiling points of its components. Naphtha is sometimes used as a diluent for W/O emulsions.

NAPL   → Nonaqueous-Phase Liquid.

Natural Organic Matter   → Total Organic Carbon.

nDMA   → Differential Mobility Analysis.

Near-Field Scanning Optical Microscopy   (NSOM) A version of scanning probe microscopy in which an optical probe, that is much smaller than the wavelength of light, is scanned over a surface while measuring the transmitted or reflected light. Thus NSOM can involve either transmission or reflectance light imaging. Also termed scanning near-field optical microscopy (SNOM). This technique can achieve resolution of the order of 100 nm, and can be used to scan quantum dots.

Neat Phase   → Neat Soap.

Neat Soap   A mesomorphic (liquid-crystal) phase of soap micelles, oriented in a lamellar structure. Neat soap contains more (e.g., 75%) soap than water. Neat soap is in contrast to middle soap, which contains less soap than water and is also a mesomorphic phase, but has a hexagonal array of cylinders rather than a lamellar structure. *See also* Reference [4].

Nebulization   → Atomization.

Negative Adsorption → Adsorption.

Negative Tactoids	$\rightarrow$ Tactoid.
Nelson-Type Emulsions	Several types of phase behavior occur in microemulsions; they are denoted as Nelson type II⁻, type II +, and type III. These designations refer to equilibrium phase behaviors and distinguish, for example, the number of phases that can be in equilibrium and the nature of the continuous phase. *See also* Reference [177]. Winsor-type emulsions are similarly identified, but with different type numbers.
Nematic State	$\rightarrow$ Mesomorphic Phase.
NEMS	$\rightarrow$ Nanoelectromechanical System.
Nephelometric Turbidity Unit	(NTU) A unit of measurement in nephelometry (turbidity). Empirical standards are available from light scattering equipment manufacturers.
Nephelometry	The study of the light-scattering properties of dispersions. In general, a nephelometer is an instrument capable of measuring light scattering by dispersions at various angles. $\rightarrow$ Light Scattering, $\rightarrow$ Turbidity.
Nernst, Walther (Hermann) (1864–1941)	A German physicist and inventor known for his work in electrochemistry and thermochemistry, including the third law of thermodynamics. He also contributed to the founding of the discipline of physical chemistry and made contributions to the theory of dissociation of ions in solution, and to understanding electric potential and the role of metal electrodes in electrochemical cells. Eponyms include the "Nernst equation" (electrochemistry). Nernst was awarded the Nobel prize in chemistry (1920) for his work in thermochemistry. *See* Reference [178].
Nested Nanoparticle	$\rightarrow$ Nano-Onion.
Neumann's Triangle	At the junction where three phases meet, three vectors representing the forces of interfacial tension among pairs of phases can be drawn. At equilibrium, the sum of these vectors of Neumann's triangle will equal zero.
Neutral Agarose	One of the kinds of polysaccharide structure comprising agar. Also termed "agaran". $\rightarrow$ Agar.
Newton Black Film	$\rightarrow$ Black Film.
Newtonian Flow	Fluid flow that obeys Newton's law of viscosity. Non-Newtonian fluids can exhibit Newtonian flow in certain shear-rate or shear-stress regimes. $\rightarrow$ Newtonian Fluid.
Newtonian Fluid	A fluid or dispersion whose rheological behavior is described by Newton's law of viscosity. Here shear stress is set proportional to shear rate. The proportionality constant is the coefficient of

viscosity, or, simply, viscosity. The viscosity of a Newtonian fluid is a constant for all shear rates.

NIL   → Nanoimprint Lithography.

NIMS   → Nanostructured Initiator Mass Spectrometry.

Niosome   → Vesicle.

Nitrified Foam   A slang term used in some industries to denote foams in which nitrogen is the gas phase.

NMR Imaging   → Magnetic Resonance Imaging.

NOM   Natural Organic Matter. → Total Organic Carbon.

Nomarski Microscopy   A kind of reflected-light microscopy in which a differential interference contrast technique is used to render a relief-like image in interference colors.

Nonaqueous-Phase Liquid   (NAPL) Any liquid other than water. In environmental fields this term commonly refers to petroleum hydrocarbons less dense than water (light nonaqueous-phase liquid, LNAPL), or oils such as chlorinated hydrocarbons, like tetrachloroethylene, that are more dense than water (dense nonaqueous-phase liquid, DNAPL).

Nondraining Polymer   Polymer molecules for which the interior of the coiled portions of the molecules are not affected by flow.

Nonionic Surfactant   A surfactant molecule whose polar group is not electrically charged. Example: poly(oxyethylene) alcohol, $C_nH_{2n+1}(OCH_2CH_2)_mOH$.

Nonionic Surfactant Vesicle Synonym for niosome. → Vesicle.

Non-Newtonian Flow   Fluid flow that does not obey Newton's law of viscosity. In the older literature non-Newtonian behaviour was sometimes referred to as anomalous [39]. Non-Newtonian fluids can exhibit non-Newtonian flow only in certain shear-rate or shear-stress regimes. A number of categories of non-Newtonian flow are distinguished, including dilatant, pseudoplastic, thixotropic, rheopectic, and rheomalaxic. → Newtonian Fluid.

Non-Newtonian Fluid   A fluid whose viscosity varies with applied shear rate (flow rate). In the older literature non-Newtonian behaviour was sometimes referred to as anomalous [39]. → Newtonian Fluid.

Non-Transitive Nanoparticle A nanoparticle that does not exhibit size-related intensive properties, that is, whose properties fall on a continuum that can be smoothly extrapolated from the behavior of the bulk (larger-scale) material [32]. Example: non-transitive nanoparticles are applied in industries that exploit their features, such as minimal optical scattering or high surface areas, to improve the radiation absorption, abrasion resistance or mechanical strength of materials. → Transitive Nanoparticle.

Nonviscous Fluid   → Inviscid Fluid.

Nonwetting   → Wetting.

Normal Photoelectron Diffraction   (NPD) → Photoelectron Diffraction.

Noüy, Pierre du   → du Noüy, Pierre (Lecomte).

NPD   Normal Photoelectron Diffraction. → Photoelectron Diffraction.

NRA   Nuclear Reaction Analysis. → Ion Beam Analysis.

NSOM   → Near-Field Scanning Optical Microscopy.

NTU   → Nephelometric Turbidity Unit.

Nuclear Magnetic Resonance Imaging   → Magnetic Resonance Imaging.

Nuclear Reaction Analysis   (NRA) → Ion Beam Analysis.

Nucleation Aerosol   An aerosol in which the primary particles or droplets have formed completely, or mostly, due to nucleation from a supersaturated vapour, and typically occur in the size range of 1 to 50 nm. → Accumulation Aerosol and → Aerosol.

Nuclei   As a solute becomes insoluble, the formation of a new phase has its origin in the formation of clusters of solute molecules, termed "germs", that increase in size to form small crystals or particles, termed "nuclei". One means of preparing colloidal dispersions involves precipitation from solution onto nuclei, which can be of the same or different chemical species. → Condensation Methods.

Number-Average Quantities   A method of averaging in which the sum of the amount of species multiplied by the property of interest is divided by the total amount of species. An example is the number- average relative molecular mass determined by osmotic pressure measurements,

$$M_{r,n} = \frac{\sum n_i M_r(i)}{\sum n_i}$$

where n_i is the amount of species and $M_r(i)$ is the relative molecular mass of species i. → Mass-Average Quantities.

Numerical Aperture   An indication of the ability of a lens to gather and transmit light.

O

Oakes Mixer	A machine used for preparing foams in the food industry. A slurry is continuously stirred and aerated under pressure between a series of blades. → Aerator.
Octanol-Water Partition Coefficient	The partitioning coefficient of a compound between octanol and water, that is, between specific nonpolar and polar phases. Used as an indication of the tendency of a compound to partition between oil and water phases. A variety of empirical equations estimate such partitioning of a compound on the basis of its octanol-water partition coefficient. → Solvent-Motivated Sorption.
Oden's Balance	An apparatus for determining sedimentation rates in which a balance pan is immersed in a sedimentation column and is used to intercept and accumulate sedimenting particles, whose mass can be determined as a function of time.
Ohnesorge Equation	An expression giving the critical velocity, V_0, needed for a liquid jet to break up into droplets and form an emulsion, as: $\eta/(\rho\gamma D)^{1/2} = 2000(\eta/[V_0\rho D])^{4/3}$, where η and ρ are the viscosity and density of the liquid in the jet, respectively; γ is the interfacial tension; and D is the nozzle diameter. It has been suggested [179] that this equation should instead be referred to as the "Richardson equation".
OHP	Outer Helmholtz plane. → Helmholtz Double Layer.
Oil	(1) An organic liquid, or (2) Liquid petroleum (sometimes including dissolved gas) that is produced from a well. In this sense oil is equivalent to crude oil. The term "oil" is, however, frequently more broadly used and can include, for example, synthetic hydrocarbon liquids, bitumen from oil (tar) sands, fractions obtained from crude oil, and liquid fats (e.g., triglycerides). → Crude Oil.

Dictionary of Nanotechnology, Colloid and Interface Science. Laurier L. Schramm
Copyright © 2008 WILEY-VCH Verlag GmbH & Co. KGaA, Weinheim
ISBN: 978-3-527-32203-9

Oil-Assisted Flotation	Any of a family of flotation processes in which oil is used to agglomerate or bridge particles and enhance flotation. In extender flotation a small amount of oil is added to improve a collector's performance. In agglomerate flotation oil is added to agglomerate very finely divided particles to a size that can be efficiently floated. In emulsion flotation oil is added to function as the collector, for already hydrophobic particles. *See* Reference [180].
Oil-Base Mud	An emulsion drilling fluid (mud) of the water-dispersed-in-oil (W/O) type having a low water content. $\rightarrow$ Oil Mud, $\rightarrow$ Invert-Oil Mud.
Oil Color	A qualitative test for the presence of emulsified water in an oil. Emulsified water droplets tend to impart a hazy appearance to the oil.
Oildag	A commercial colloidal graphite product (a suspension of colloidal graphite in oil). $\rightarrow$ Acheson, Edward.
Oil Emulsion	An emulsion having an oil as the continuous phase.
Oil-Emulsion Mud	An emulsion drilling fluid (mud) of the oil-dispersed-in-water (O/W) type. $\rightarrow$ Oil Mud.
Oil Hydrosol	An oil-in-water (O/W) emulsion in which the oil droplets are very small and the volume fraction of oil is also very small. The emulsion terminology is preferable. $\rightarrow$ Hydrosol.
Oil Mud	An emulsion drilling fluid (mud) of the water-dispersed-in-oil (W/O) type. A mud of low water content is referred to as an "oil-base mud", and a mud of high water content is referred to as an "invert-oil mud". $\rightarrow$ Oil-Emulsion Mud.
Oleomargarine	$\rightarrow$ Margarine.
Oleo Oil	$\rightarrow$ Margarine
Oleophilic	The oil-preferring nature of a species. A synonym for lipophilic. $\rightarrow$ Hydrophobic.
Oleophobic	The oil-avoiding nature of a species. A synonym for lipophobic. $\rightarrow$ Hydrophilic.
Oligomer	A relatively short-chain polymer, typically having a degree of polymerization of less than about 10.
Oliver-Ward Equation	An empirical equation for estimating the viscosity of a dispersion. *See* Table 11.
One-Second Criterion	A rule of thumb related to the degree of suspension stability needed for successful mineral flotation. It states that the particles in a suspension are sufficiently well dispersed for flotation if

individual particles do not remain settled at the bottom of the flotation vessel for longer than one second.

Onion Phase

The lyotropic lamellar phase state in which the lamellae organize themselves into a collection of multilamellar vesicles. *See* Reference [181].

Onsager, Lars (1903–1976)

A Norwegian-born American chemical engineer and physical chemist known for his work in electrolyte solution theory, as well as for his contributions in diffusion, colloidal solutions, and turbulence. He made an improvement to the Debye-Hückel theory of electrolyte solutions, known as the *Onsager Limiting Law*. He won the Nobel prize in chemistry (1968) for the discovery of *Onsager's Reciprocal Relations*, which are fundamental for the thermodynamics of irreversible processes, and have been termed the "4th law of thermodynamics". *See* Reference [182].

O/O

Abbreviation for an oil-dispersed-in-oil emulsion in which one oil is polar and the other is not. Example: an emulsion of ethylene glycol in a liquid alkane.

Opacifiers

Agents that make a liquid appear more opaque, or pearlescent. For example, polystyrene latex is added to liquid detergents formulated for dishwashing or shampooing to give them a flat opaque appearance. → Detergent.

Open-Cell Foam

→ Closed-Cell Foam.

Optical Brighteners

Agents that make treated materials appear more white. For example, fluorescent whitening agents are added to laundry detergents so that they can become attached to fibres and give an enhanced whiteness by absorbing UV light and emitting blue light. Example: stilbene disulfonates. → Detergent.

Optical Lever

→ Scanning Probe Microscopy.

Optically Stimulated Exoelectron Emission Spectroscopy

(OSEE) A means of determining the presence and nature of adsorbed species by measuring the electron emission caused by having a beam of light strike a surface in the presence of an electric potential field. *See also* Table 7.

Optical Second Harmonic Generation

(SHG) A laser technique based on a second-order nonlinear process that occurs at a surface (where there are discontinuities) but not in the bulk phases. An optical beam from a laser, of frequency ω, is made to strike a surface at an angle. The beam induces dipole oscillation of molecules. Radiation from molecules at the surface, at the second-harmonic frequency 2ω, is collected and used for monitoring surface dynamics and surface reactions. *See also* References [183, 184].

Optical Tweezing Electrophoresis	A method for single-particle microelectrophoresis using optical tweezers to trap a particle and then applying an alternating electric field to create the electrophoretic motion. This has been applied to extremely dilute dispersions for which laser Doppler velocimetry, for example, could not be applied. *See* Reference [185]. → Optical Tweezers.
Optical Tweezers	In microscopy, optical tweezers refers to the use of the electric field gradients caused by focussed beams of laser light to trap and position particles having a higher dielectric constant than the dispersing phase. These particles are trapped in the "waist" of the converging laser beam and are forced to move as the waist is made to move. The same laser beam is used to image the trapped particle. [186] → Nanotweezers, → Optical Tweezing Electrophoresis.
Optimum Salinity	In microemulsions, the salinity for which the mixing of oil with a surfactant solution produces a middle-phase microemulsion containing an oil-to-water ratio of 1. In micellar enhanced oil recovery processes, extremely low interfacial tensions result, and oil recovery tends to be maximized when this condition is satisfied.
Organosol	A dispersion of very small diameter species in an organic medium. → Hydrosol.
Orientation Forces	Keesom forces. → van der Waals Forces.
Oriented-Wedge Theory	An empirical generalization used to predict which phase in an emulsion will be continuous and which dispersed. It is based on a physical picture in which emulsifiers are considered to have a wedge shape and will favor adsorbing at an interface such that most efficient packing is obtained; that is, with the narrow ends pointed toward the centers of the droplets. A useful starting point, but there are many exceptions. → Bancroft's Rule, → Hydrophile-Lipophile Balance.
Orimulsion®	A commercial emulsion fuel for power-generating plants made from bitumen that comes from the Orinoco River deposit in Venezuela. Orimulsion® is an O/W emulsion, containing about 70% oil. Being water-continuous the emulsion is easily handled and transported, and otherwise behaves similarly to fuel oil.
Orogenic Displacement	The displacement of proteins from an interface by introducing a separate surfactant that adsorbs where there are defects in the interfacial protein network. Increasing adsorption of the surfactant exerts a surface pressure that acts to compress the protein network. At sufficiently high pressure the protein network fails, releasing proteins that desorb. *See* Reference [187].

Orthokinetic Aggregation	The process of aggregation induced by hydrodynamic motions such as stirring, sedimentation, or convection. Orthokinetic aggregation is distinguished from perikinetic aggregation, the latter being caused by Brownian motions.
Orthowater	→ Polywater.
Oscillating Jet Method	A method for the determination of surface or interfacial tension in which a stream of gas or liquid is injected into another liquid phase through a jet having an elliptical orifice. Oscillations develop in the jet. The jet dimensions and wavelength are measured and used to calculate the surface or interfacial tension. Also termed the "elliptical jet method".
OSEE	→ Optically Stimulated Exoelectron Emission Spectroscopy.
Osmometer	An instrument for determining the osmotic pressure exerted by solvent molecules diffusing through a semipermeable membrane in contact with a solution or hydrophilic colloidal dispersion. → Colloid Osmotic Pressure, → Osmotic Pressure.
Osmosis	The process in which solvent will flow through a semipermeable membrane (permeable to solvent but not to solute) from a solution of lower dissolved solute activity (concentration) to that of higher activity (concentration). → Donnan Equilibrium, → Osmotic Pressure.
Osmotic Pressure	When a solution of dissolved species is separated from pure solvent by a semipermeable membrane, not permeable to the dissolved species, the osmotic pressure is the pressure difference required to prevent transfer of the solvent; that is, to prevent osmosis. → Colloid Osmotic Pressure.
Ostwald, (Carl Wilhelm) Wolfgang (1883–1943)	Originally trained as a zoologist, the Latvian-born German Ostwald became a founder of colloid chemistry and originated the basic definition of colloidal dispersions. He worked in a variety of colloid chemical areas, including rheology, precipitation, coagulation, and foams. Eponyms include "Ostwald ripening" (crystal growth from solution) and "Ostwald's law" (conductivity and dissociation in weak acids and bases). The founder of Kolloid Zeitschrift (now Colloid & Polymer Science) in 1906, he also published five books in the field, including *Die Welt der vernachlässigten Dimensionen* (*The World of Neglected Dimensions*) in 1914. His father Wilhelm Ostwald was a founder of the broader field of physical chemistry. *See* Reference [188].
Ostwald, (Friedrich) Wilhelm (1853–1932)	A Latvian-born German chemist who (with van't Hoff and Arrhenius) helped establish the discipline of physical chemistry. He made contributions in many areas, including chemical

equilibria, electrolytic dissociation, crystallization, and electrical conductivity. He wrote many textbooks and organized the first journal of physical chemistry, Zeitschrift für physikalische Chemie, in 1887. He received the Nobel prize in chemistry (1909) for his work on catalysis and on the principles governing chemical equilibria and rates of reaction. His son Wolfgang Ostwald was a famous colloid chemist. *See* References [189, 190].

Ostwald Ripening	The process by which larger droplets or particles grow in size in preference to smaller droplets or particles because of their different chemical potentials. → Aging.
Ostwald Viscometer	→ Capillary Viscometer.
Outer Helmholtz Plane	(OHP) → Helmholtz Double Layer.
Outer Potential	The potential just outside the interface bounding a specified phase. Also termed the "Volta potential". The difference in outer (Volta) potentials between two phases in contact is equal to the surface or interfacial potential between them. → Inner Potential, → Jump Potential.
Outgassing	The desorption of gas from a solid under conditions of extremely high vacuum and temperature.
Overbeek, (Jan) Theodoor (Gerard) (1911–2007)	A Dutch physical chemist known for his work in colloidal dispersions and for numerous books on the subject. Together with E.J.W. Verwey he developed a theory of the stability of colloidal particles that could explain the Schulze-Hardy rule. Independently developed also by Derjaguin and Landau the theory became famous as the "DLVO theory", an acronym constructed from the last names of these four scientists.
Overpotential	The additional electrical potential beyond that of the thermodynamic electrode potential required to cause current to flow in an electrochemical cell. Also termed "overvoltage".
Overrun	A measure of foaming capacity used in the food industry to evaluate the effectiveness of proteins and other stabilizing agents. The overrun is related to the amount of interfacial area stabilized by the foaming surfactant and is given by $100 \, (V_F - V_L)/V_L$, where V_F is the foam volume and V_L is the volume of initial liquid.
Overvoltage	→ Overpotential.
O/W	Abbreviation for an oil-dispersed-in-water emulsion.
O/W/A	Abbreviation for a fluid film of water between oil and air. → Fluid Film.

O/W/O

(1) In multiple emulsions: Abbreviation for an oil-dispersed-in-water-dispersed-in-oil multiple emulsion. The water droplets have oil droplets dispersed within them, and the water droplets themselves are dispersed in oil forming the continuous phase.

(2) In fluid films: Abbreviation for a thin fluid film of water in an oil phase. Not to be confused with the multiple emulsion convention.
$\rightarrow$ Fluid Film.

P

Pad Layer Emulsion	$\rightarrow$ Interface Emulsion.
Palisade Layer	In a micelle, the region of water molecules of hydration postulated to lie between relatively water-free hydrocarbon chains at the center of the micelle and the exposed, fully hydrated polar groups at the micelle surface.
Pallmann Effect	$\rightarrow$ Suspension Effect.
PALM	Photoactivated Localization Microscopy. $\rightarrow$ Super-Resolution Microscopy.
Paracasein	$\rightarrow$ Casein.
Parachor	An empirical parameter used in the estimation of the surface tension of liquids. The parachor, $P = M\gamma^{1/4}/\Delta\rho$ where γ is the surface or interfacial tension, $\Delta\rho$ is the density difference between the phases, and M is the molecular mass of the liquid. *See* Table 8.
Paraffin-chain Salts	An older term for "ionic surfactants".
Paramagnetic	A material that acquires magnetic properties when placed in an external magnetic field and becomes attracted into the magnetic field (the attraction is proportional to the strength of the applied field). In contrast, diamagnetic materials are weakly repelled by an external magnetic field, and ferromagnetic materials have magnetic properties independent of external magnetic fields. $\rightarrow$ Ferromagnetic.
Particle	A small piece, or fragment, of a solid material. Example: a particle of sand in a beach.
Particle Induced Gamma-Ray Emission Analysis	(PIGEA) $\rightarrow$ Ion Beam Analysis.

Dictionary of Nanotechnology, Colloid and Interface Science. Laurier L. Schramm
Copyright © 2008 WILEY-VCH Verlag GmbH & Co. KGaA, Weinheim
ISBN: 978-3-527-32203-9

Particle Induced X-ray Emission Analysis	(PIXEA) → Ion Beam Analysis.
Particle Microelectrophoresis	→ Electrophoresis.
Particle Size	The size of a particle as determined by a particular method. It is prefereable to specify the measurement method used.
Particle Size Classification	The separation or determination of particles into different size ranges. A number of classification systems are used, some of which correspond to physical means of separations such as by sieves. Examples are given in Table 12.
PAS	→ Photoacoustic Spectroscopy.
Pascalian Fluid	→ Inviscid Fluid.
Passivation	The process by which a nonconducting crystal layer is caused to grow on the surface of a conductor or semiconductor surface to protect the surface from electrical conduction and chemical reaction (e.g., corrosion).
Paucidisperse	A colloidal dispersion in which the dispersed species (droplets, particles) have a few different sizes. Paucidisperse is a category of heterodisperse systems. → Monodisperse.
PBG	→ Photonic Band Gap Structure.
PCS	→ Photon Correlation Spectroscopy.
Peel Test	A means of determining the strength of a joint between two materials. The force normal to the surface required to separate the joint is measured (adhesion), or, alternatively, the force parallel to the surface needed to separate the joint (shear adhesion) is measured.
PEEM	Photoemission Electron Microscopy. → Photoelectron Spectroscopy.
Pendant Bubble Method	→ Pendant Drop Method.
Pendant Drop Method	A method for determining surface or interfacial tension based on measuring the shape of a droplet hanging from the tip of a capillary (in interfacial tension the droplet can alternatively hang upwards from the tip of an inverted capillary). Also termed the "hanging drop (or bubble) method".
Peptization	The dispersion of an aggregated (coagulated or flocculated) system. Same meaning as Deflocculation.
Peptizing Ions	→ Potential-Determining Ions.
Percolation	(1) The slow movement of a liquid through a porous medium. Example: the downward movement of a dense liquid in a soil environment.

(2) A condition in a dispersed system in which a property such as conductivity increases strongly at a critical concentration, termed the "percolation threshold", as a result of the formation of continuous conducting paths, termed "infinite clusters".

Percolation Threshold	$\rightarrow$ Percolation.
Perikinetic Aggregation	The process of aggregation when induced by Brownian motions. Perikinetic aggregation is distinguished from orthokinetic aggregation, the latter is caused by hydrodynamic motions such as sedimentation or convection.
Permeability	A measure of the ease with which a fluid can flow (fluid conductivity) through a porous medium. Permeability is defined by Darcy's law. For linear, horizontal, isothermal flow, permeability is the constant of proportionality between flow rate times viscosity and the product of cross-sectional area of the medium and pressure gradient along the medium.
Permeate	$\rightarrow$ Dialysis.
Permittivity	A measure of the ability of a medium to affect an applied electric field. It is the constant of proportionality between the force acting between two point charges and the product of the two charges divided by the square of the distance separating them. The relative permittivity of a material equals the permittivity of the material multiplied by that of vacuum. An older term for relative permittivity is "dielectric constant". *See also* Table 9.
Perrin Black Film	A Newton black film. $\rightarrow$ Black Film.
Perrin, Jean (Baptiste) (1870–1942)	A French physicist and physical chemist known for contributions to cathode ray physics, he is known to colloid science for experimental studies of sedimentation and Brownian motions, the latter of which proved the existence of atoms and molecules. He may have introduced the term "hydrophobic". Certain thin liquid films that appear black when illuminated by reflected white light were at one time known as "Perrin black films" (now more commonly called "Newton black films"). Perrin won the Nobel prize (1926) in physics for experiments demonstrating the existence of molecules, and for his discovery of sedimentation equilibrium. *See* References [191, 192].
Persorption	Selective adsorption of small molecules rather than large molecules in the pores of an adsorbent attributable to the small sizes of the pores. For such an adsorbent, the amount of adsorption varies with the molecular size of the adsorbate. $\rightarrow$ Molecular Sieve Effect.
PES	$\rightarrow$ Photoelectron Spectroscopy.

Petroleum	A general term that can refer to any natural source, geologically sourced hydrocarbons or hydrocarbon mixtures, usually liquid, but sometimes solid or gaseous.
Phase Diagram	A graphical representation of the equilibrium relationships between phases in a system. For multicomponent systems, and considering varying temperatures, more than a simple two-dimensional phase diagram will be required.
Phase Inversion Temperature	(PIT) The temperature at which the hydrophilic and oleophilic natures of a surfactant are in balance. As temperature is increased through the PIT, a surfactant will change from promoting one kind of emulsion, such as O/W, to another, such as W/O. Also termed the "HLB temperature".
Phase Map	→ Phase Diagram.
Phase Ratio	In emulsions, the term refers to the ratio between internal phase and continuous phase. Phase ratios are dimensionless, but the units used should be specified because mass ratios and volume ratios are commonly used.
Phase Rule	The fundamental thermodynamic equation governing the equilibria between phases in a system. Also termed the "Gibbs phase rule", it specifies the number of intensive variables needed to describe a system (degrees of freedom, f) in terms of the numbers of components, c, and phases, p, present as $f = c - p + 2$.
Phase-Transfer Catalysis	A catalytic reaction in which a surfactant is used to form an ion-pair with a water-soluble reactant and aid the transport of the reactant into an organic phase, where it reacts with other reactant(s) soluble only in the organic phase. → Micellar Catalysis.
PhD	→ Photoelectron Diffraction.
Phospholipid	Esters of phosphoric acid that contain fatty acid(s), an alcohol, and a nitrogen-containing base. → Lipid.
Phospholipid Bilayer	→ Bimolecular Film.
Photoacoustic Spectroscopy	(PAS) A technique for studying the vibrational states of surface species and species adsorbed on surfaces. A sample is placed inside a gas-containing chamber and irradiated with radiation of a given wavelength having an intensity modulated at an acoustic frequency. Absorption of this modulated radiation causes periodic heat flow from the sample and generates sound waves that are detected. *See also* Table 7.
Photoactivated Localization Microscopy	(PALM) → Super-Resolution Microscopy.
Photoelectric Effect	When photons interact with bound electrons in atoms or molecules causing the production of photoelectrons and excited ions.

Photoelectron Diffraction	(PhD) A diffraction technique similar to high-energy electron diffraction except that a higher energy (X-ray) beam is used at shallow angles to cause the ejection of photoelectrons. Again, the diffraction pattern yields information about surface structure. The distinctions "normal photoelectron diffraction (NPD)" and "azimuthal photoelectron diffraction (APD)" are sometimes used. *See also* Table 7.
Photoelectron Spectroscopy	(PES) A technique related to Auger electron spectroscopy and also used for the determination of surface composition. The surface is scanned with a photon beam causing the ejection of electrons from the surface atoms. The energies of the ejected electrons are determined (both primary photoelectrons and the Auger electrons); these energies are characteristic of the atoms from which they were ejected. The principal types of photoelectron spectroscopy use different photon sources: ultraviolet photoelectron spectroscopy (UPS) and X-ray photoelectron spectroscopy (XPS). When XPS was originally developed for chemical analysis the technique was referred to as "electron spectroscopy for chemical analysis (ESCA)", although the term "XPS" is mostly used now because other electron spectroscopic techniques can also be used for chemical information. XPS is also termed photo emission electron microscopy (PEEM). → Auger Electron Spectroscopy, *see* Table 7. *See* Reference [57] for specific terms in photoelectron spectroscopy.
Photoemission Electron Microscopy	(PEEM) → Photoelectron Spectroscopy.
Photographic Emulsion	Not an emulsion but rather a dispersion (solid suspension) of silver halide particles in gelatin.
Photomicrograph	A photographic image formed by a microscope. The resulting photographic image is much larger than the original object being photographed. This image is not the same as a microphotograph.
Photon Correlation Spectroscopy	(PCS) A means of determining particle size; scattered photons are correlated with the microscopic (Brownian) motion of particles suspended in a fluid. The time fluctuation of the scattering is related to the diffusion of the particles in the medium, and therefore to their size. Synonyms include dynamic light scattering (DLS), quasi-elastic light scattering (QELS) and self-beat light scattering.
Photonic Band Gap Structure	(PBG) Nanosized structures, such as a multilayered stack, designed to manipulate the properties of light by modifying the relative permittivity of the layers. PBGs are of interest in the construction of devices for sensing and chemical characterization. *See* Reference [37].

Photon-Stimulated Desorption Spectroscopy	(PSD) A surface technique in which electronically stimulated adsorbed species are desorbed from a surface to gain information about adsorbate-substrate bonding and about surface composition. In photon-stimulated desorption, photon bombardment is used to excite the adsorbed species, whereas in electron-stimulated desorption (ESD), low-energy electrons (less than 500 eV) are used. $\rightarrow$ Temperature-Programmed Reaction Spectroscopy.
Photophoresis	Particle motion, in an aerosol of solid particles, caused by non-uniform heating due to absorption in the particle of radiation. The particles will tend to move parallel to the direction of the light. $\rightarrow$ Thermophoresis.
Photozone Counter	A particle- or droplet-sizing technique, analogous to the electrical sensing-zone methods, that relies on visible light absorption in sample introduced into a small chamber. The particles or droplets must be greater than the wavelength of the light used, to minimize scattering (>1 μm). $\rightarrow$ Sensing-Zone Technique.
Physical Adsorption	$\rightarrow$ Chemisorption.
Physical Vapour Deposition	(PVD) A gas-phase particle synthesis method in which one or more of the feedstock materials goes through a gas phase, nucleation on a surface site, and then growth through condensation and/or coalescence. Example: PVD is used to produce metalized polyethylene terephthalate polyester films in "foil" food packaging. PVD is also used to make Sculptured Thin Films. $\rightarrow$ Chemical Vapour Deposition.
Physisorption	$\rightarrow$ Chemisorption.
Pickering Emulsion	An emulsion stabilized by fine particles. The particles form a close-packed structure at the oil-water interface, with significant mechanical strength, which provides a barrier to coalescence.
Picofluidics	The study of the behaviour of, and the control or manipulation of fluids on the picolitre scale. $\rightarrow$ Microfluidics, $\rightarrow$ Biomedical Microelectromechanical Systems, $\rightarrow$ Microelectromechanical Systems, $\rightarrow$ Micrototal Analysis System, $\rightarrow$ Nanoelectromechanical System.
Picoscale	Having at least one dimension in the range 1 to 100 pm, where a picometre (pm) $= 10^{-12}$ m. Also termed Picosize.
Picosize	$\rightarrow$ Picoscale.
Picotechnology	The design, characterization, manipulation, incorporation, and/or production of materials and structures in the picoscale range, by any of a variety of physical and chemical methods. The picoscale range is typically 1 to 100 pm, where a picometre (pm) $= 10^{-12}$ m.

An important distinction is that these applications exploit the properties of the picoscale components, as distinct from bulk or macroscopic systems. There is a substantial overlap of scale between picotechnology and nanotechnology, particularly where it comes to molecular nanotechnology: the precise, controlled assembly of structures up from the molecular scale that are well-organized, and with reproducible properties. → Collodial, → Microtechnology, → Nanoscience, → Nanotechnology.

PIGEA

Particle Induced Gamma-Ray Emission Analysis. → Ion Beam Analysis.

Pigment

Insoluble material that is finely divided, micronized (for example), and uniformly dispersed in a formulated system for the purpose of coloring it or making it opaque. Examples: TiO_2 in soap bars and paints; iron oxides in eye makeup and paints.

Pigment Grind

Pigment particles dispersed in a liquid, such as castor oil. → Roll Mill.

PILC

→ Pillar Interlayered Clay Minerals.

Pillar Interlayered Clay Minerals

Clay mineral particles frequently carry a significant electrical charge, which is compensated for by counterions. If the counterions are very large then those present between clay layers cause a significant interlayer separation. Such materials containing an intercalation layer are termed pillar interlayered clay minerals, pillared inorganic layered compounds (PILC), or composite clay nanostructures. → Intercalation.

Pin Mill

Also termed Toothed Disc Mill, a kind of rotor-stator homogenizer. The force for disruption comes from the energy of the rotor and the shearing action arises from the narrow gap between rotor and stator. → Colloid Mill.

PIT

→ Phase Inversion Temperature.

PIXEA

Particle Induced X-ray Emission Analysis. → Ion Beam Analysis.

Plait Point

In phase diagrams the composition condition for which three coexisting phases, containing partially soluble components, of a three-phase system all approach the same composition.

Plasmalemma

→ Cell Membrane.

Plasma Membrane

→ Cell Membrane.

Plasma Processing

A gas-phase fine particle preparation method in which a plasma reactor is used to cause evaporation or induce chemical reactions in the gas phase. The reaction product goes through nucleation and then growth through condensation, coagulation, and/or

	coalescence. Different kinds of plasma can be used, including arc, direct-current, and radio-frequency plasmas.
Plastic Flow	The deformation or flow of a solid under the influence of an applied shear stress.
Plastic Fluid	A fluid characterized by both of the following: the existence of a finite shear stress that must be applied before flow begins (yield stress), and Newtonian flow at higher shear stresses. May be referred to as "Bingham plastic". $\rightarrow$ Generalized Plastic Fluid.
Plasticity Index	$\rightarrow$ Plasticity Number.
Plasticity Number	The difference between the liquid limit and the plasticity limit of a soil or similar material [43, 44]. Also termed the "plasticity index". $\rightarrow$ Atterberg Limits, $\rightarrow$ Liquid Limit, $\rightarrow$ Plastic Limit.
Plasticizer	(1) General. Any additive that contributes flexibility when added into a formulated product. Plasticizers usually permit less water to be used in a formulation. There is a wide range of applications, two of which are described below. Also termed flexibilizer.
	(2) Polymers. An additive used in polymer (usually polyvinyl chloride) formulations to help make them more flexible and stretchable, less brittle, and also to improve the polymer's ability to be molded and thermally cured to hold its molded shape. Phthalates are the most common plasticizers, especially dioctyl phthalate which is used for polyvinyl chloride.
	(3) Concrete. An additive used in concrete formulations to reduce their water-content requirement while maintaining workability. Typical concrete placticizers are based on carboxylic or sulphonic acids, such as lignosulphonic acid. Also termed water-reducers, fluidizers, or fluidifiers. Superplasticizers are similar to plasticizers except that they are capable of reducing the water-content requirements by up to 30 per cent compared with the 10–15 percent that is typical of conventional concrete plasticizers. Typical concrete superplasticizers include sulphonated melamine-formaldehyde condensates (SMF) and sulphonated naphthalene-formaldehyde condensates (SNF), among others. Superplasticizers are also termed high-range water-reducers, superfluidizers, superfluidifiers, or super water-reducers.
Plastic Limit	The minimum water content for which a small sample of soil or similar material will barely deform or crumble in a standardized test method [43, 44]. Also termed the "lower plastic limit". $\rightarrow$ Atterberg Limits, $\rightarrow$ Liquid Limit, $\rightarrow$ Plasticity Number.
Plastisol	A solution of polyvinyl chloride (PVC) polymer resin in a liquid containing a plasticizer. The plasticizer additive helps make them

more flexible and stretchable, and also improves the polymer's ability to be molded and thermally cured to hold its molded shape. The first plastisol products were replacements for natural rubber in wire insulation and molded protective covers for machine parts and tools. Plastisol pastes are also used in the formulation of painting, casting, and spraying of protective coatings, and in the formulation of plastisol inks.

Plateau Border	The region of transition at which thin fluid films are connected to other thin films or mechanical supports such as solid surfaces. For example, in foams plateau borders form the regions of liquid situated at the junction of liquid lamellae. Sometimes referred to as a "Gibbs ring" or "Gibbs-Plateau Border".
Platelet-Nanofibre	Nanofilaments built from graphenes oriented perpendicular to the filament axis and resembling stacked plates. These are carbon multi-wall nanotubes (MWNT) and represent an example of h-MWNT except that there is no open, inner cavity. $\rightarrow$ Carbon Nanotube, $\rightarrow$ Nanofibre.
PLAWM Trough	Pockels-Langmuir-Adam-Wilson-McBain trough. $\rightarrow$ Film Balance.
Plumbago	$\rightarrow$ Graphite.
Pockels, Agnes (1862–1935)	A self-taught Austrian chemist who, working in her home kitchen, made important contributions in the areas of surface films and surface tension and can be regarded as a founder of surface chemistry. She developed sample preparation techniques for ensuring clean surfaces and a "Pockels' slide trough" or film balance technique for the study of surface films, which was later extended by Langmuir and others, and is sometimes known as a "PLAWM (Pockels-Langmuir-Adam-Wilson-McBain) trough". She calculated the thickness of monomolecular surface films and the areas per molecule in the films; the corresponding minimum area of ca. $20\,\text{Å}^2$/molecule is known as the "Pockels' point". *See* References [193–195].
Pockels Effect	The change in refractive properties under the influence of an applied electric field that occurs for some crystals. (Named for F. Pockels).
Pockels-Langmuir-Adam-Wilson-McBain Trough	$\rightarrow$ Film Balance.
Pockels Point	When surfactant molecules are added into a system and form an insoluble film at an interface, surface tension does not decrease very strongly until enough surfactant is added to form a complete monolayer. The transition point is termed the "Pockels point" and corresponds to a surface area occupied per molecule of about $20\,\text{Å}^2$ for soaps. (Named for A. Pockels).

Point of Zero Charge	The condition, usually the solution pH, at which a particle or interface is electrically neutral. This condition is not always the same as the isoelectric point, which refers to zero charge at the shear plane that exists a small distance away from the interface. → Electrocapillarity, → Colloid Titration.
Poiseuille Flow	→ Hagen-Poiseuille Law.
Poiseuille, Jean Louis (Marie) (1799–1869)	A French physiologist who was probably the first to quantitatively study viscosity. Poiseuille developed an improved way to measure dynamic blood pressures using a mercury manometer whose connection to an artery was filled with with potassium carbonate to prevent coagulation (named the hemodynamometer). As a physician he was interested in the movement of blood through blood vessels, but his observations proved to more general than this. Poiseuille's law describing the volumetric flow rate for liquids through small pipes and capillaries was initially named after him, but later renamed as the Hagen-Poiseuille law in recognition of Hagen's earlier work on the same topic. The unit of viscosity Poise is named for Poiseuille.
Poison	In catalysis, any substance that interacts with a catalyst and thereby causes a reduction in catalytic activity, even when present in small concentration. Example: Trace sulfur can poison platinum-based catalysts.
Poisson-Boltzmann Equation	A fundamental equation describing the distribution of electric potential around a charged species or surface. The local variation in electric-field strength at any distance from the surface is given by the Poisson equation, and the local concentration of ions corresponding to the electric-field strength at each position in an electric double layer is given by the Boltzmann equation. The Poisson-Boltzmann equation can be combined with Debye-Hückel theory to yield a simplified, and much used, relation between potential and distance into the diffuse double layer.
Poisson Equation	A fundamental equation describing the reduction in electric-field strength that occurs with increasing distance away from a charged species in a dielectric medium. Electric-double-layer theory involves an assumption that the charge distribution is a continuous function of distance away from a charged surface, with the result that the effects of the various ion charges are averaged into layers. The Poisson equation gives the relationship between the volume charge density at a point in solution and the potential at that same point. This equation can be combined with the Boltzmann equation and Debye-Hückel theory to yield a simplified, and much used, relation between potential and distance into the diffuse double layer.

Poisson, Siméon-Denis (1781–1840)	A French mathematician, astronomer, and physicist known for his work applying mathematical principles to problems in physics, especially in mechanics. He is best known in colloid science for his work in probability theory. The local variation in electric-field strength at any distance from a charged surface is given by the Poisson equation, which is a component of the Poisson-Boltzmann Equation. He is also known for Poisson's ratio in elasticity, and for his 1833 *Treatise on Mechanics* and his 1835 *Mathematical Theory of Heat*.
Polanyi Isotherm	An adsorption isotherm equation that allows for multilayer adsorption and treats adsorbing species as though they fall into a potential energy minimum. The adsorbed layer is most compressed near the surface and becomes progressively less dense with distance away from the surface. $\rightarrow$ Adsorption Isotherm.
Polar Group	$\rightarrow$ Head Group.
Polar Substance	A substance having different, usually opposite, characteristics at two locations within it. Example: a permanent dipole. Increasing polarity generally increases solubility in water.
Polyacid	$\rightarrow$ Polyelectrolyte.
Polyampholyte	$\rightarrow$ Polyelectrolyte.
Polyanion	$\rightarrow$ Polyelectrolyte.
Polybase	$\rightarrow$ Polyelectrolyte.
Polycation	$\rightarrow$ Polyelectrolyte.
Polydisperse	A colloidal dispersion in which the dispersed species (droplets, particles) have a wide range of sizes. Polydisperse is a category of heterodisperse systems. $\rightarrow$ Monodisperse.
Polyederschaum	$\rightarrow$ Foam.
Polyelectrolyte	A kind of colloidal electrolyte consisting of a macromolecule that, when dissolved, dissociates to yield a polyionic parent macromolecule and its corresponding counterions. Also termed "polyion", "polycation", or "polyanion". Similarly, a polyelectrolyte can be referred to in certain circumstances as a "polyacid", "polybase", "polysalt", or "polyampholyte" [4]. Example: Carboxymethylcellulose.
Polyion	$\rightarrow$ Polyelectrolyte.
Polymer	A substance composed of polymer molecules (macromolecules), which are molecules made up of many repeating units, or groups, of atoms. Sometimes termed "homopolymer" to distinguish from copolymers such as block copolymers. $\rightarrow$ Associative Polymers, $\rightarrow$ Block Copolymer, $\rightarrow$ Atactic Polymer, $\rightarrow$ Tacticity.

Polymer Bridging	A mechanism of aggregation or flocculation in which long-chain polymers adsorb onto particle surfaces leaving loops and ends extending out into solution. If these loops and ends contact and adsorb onto another particle then a so-called bridge is formed. Also called bridging flocculation. *See* Reference [196]. → Aggregation.
Polymer Brush	An assembly of polymer molecules bonded at one end, usually by covalent bonds, to a surface or interface. This allows a range of surface patterns and properties to be created. The polymers act like a "brush" only when the number of polymer molecules per unit area is high enough to force the chains to adopt elongated configurations. Example: Polymer brush patterns can be used to selectively prevent adhesion.
Polymer Charge Patch	A mechanism of aggregation or flocculation in which polymers adsorb onto particles in patches because their charge density is greater than that of the surface, so complete surface coverage is not needed to neutralize the particle charge. Thus regions of positive and negative charge would exist on the particle surfaces, allowing electrostatic attraction between particles to promote aggregation. *See* Reference [136]. → Aggregation.
Polymer Colloid	A dispersion of colloidal size polymer particles in a non-solvent medium. Example: Submicroscopic latex spheres, prepared either by emulsion polymerization or by seeded emulsion polymerization, are used for a variety of calibration purposes in colloid science. *See* Reference [197].
Polymer Nanocomposite	→ Nanocomposite.
Polymer Overlap Concentration	→ Critical Aggregation Concentration.
Polymer-Thickened Foam	A foam that, in addition to the stabilizing surfactants, contains polymer. Polymer-thickened foams are formulated to produce increased stability and viscosity. → Gel Foam, → Stiff Foam.
Polymorphism	Referring to different crystal structures of the same compound. Allotropy refers to different crystal structures of the same element.
Polysalt	→ Polyelectrolyte.
Polysoap	Molecules in which surfactant monomers are incorporated into polymer chains. → Hydrophobically Associating Polymers.
Polywater	A once-postulated form of water having unusually high viscosity, high surface tension, and low vapor pressure, among other properties. Also termed anomalous water, water II, orthowater, superwater, and cyclimetric water. By 1974 it was determined that polywater does *not* exist, and that the original experimental observations were attributable to other causes. *See* the review by Adamson [5].

Pore	In porous media, the interconnecting channels forming a continuous passage through the medium are made up of pores, or openings, which can be of different sizes. Macropores have diameters greater than about 50 nm. Mesopores have diameters between about 2 and 50 nm. Micropores have diameters of less than about 2 nm. Nanopores, having diameters between 1 and 100 nm, span all three of these ranges. $\rightarrow$ Nanoporous, $\rightarrow$ Porous Medium, $\rightarrow$ Void, *see* Table 18.
Porosimeter	An instrument for the determination of pore size distribution by measuring the pressure needed to force liquid into a porous medium and applying the Young-Laplace equation. If the surface tension and contact angle appropriate to the injected liquid are known, pore dimensions can be calculated. A common liquid for this purpose is mercury, hence the term "mercury porosimetry".
Porosity	The ratio of the volume of all void spaces to total volume in a porous medium; usually expressed as a percentage. In geology "primary porosity" refers to initial, or unweathered, media, and "secondary porosity" refers to that associated with weathered media. Also termed Total Porosity. The Effective Porosity is the ratio of the volume of all interconnected void spaces (interstices) to total volume in a porous medium; usually expressed as a percentage. $\rightarrow$ Void Ratio.
Porous Medium	A solid containing voids or pore spaces. Normally such pores are quite small compared with the size of the solid and are well-distributed throughout the solid. In geologic formations porosity can be associated with unconsolidated (uncemented) materials, such as sand, or a consolidated material, such as sandstone. *See* Table 18.
Positional Synthesis	$\rightarrow$ Assembler.
Potential-Determining Ions	Ions whose equilibrium between two phases, frequently between an aqueous solution and a surface or interface, determines the difference in electrical potential between the phases, or at the surface. Example: For the AgI surface in water both Ag^+ and I^- would be potential-determining ions. If such ions are responsible for the stabilization of a colloidal dispersion, they are referred to as "peptizing ions". $\rightarrow$ Indifferent Electrolyte.
Potential Energy Diagram	$\rightarrow$ Gibbs Energy of Interaction.
Potential Energy Barrier	$\rightarrow$ Activation Energy.
Potential Energy of Attraction	$\rightarrow$ Gibbs Energy of Attraction.
Potential Energy of Interaction	$\rightarrow$ Gibbs Energy of Interaction.

Potential Energy of Repulsion	→ Gibbs Energy of Repulsion.
Pour Point	The lowest temperature at which an emulsion, oil, surfactant solution, or other material will flow under a standardized set of test conditions.
Power-Law Fluid	A fluid or dispersion whose rheological behavior is reasonably well described by the power-law equation. Here, shear stress is set proportional to the shear rate raised to an exponent n, where n is the power-law index. The fluid is pseudoplastic for $n < 1$, Newtonian for $n = 1$, and dilatant for $n > 1$. For this example and other models *see* References [21, 28].
Pressure Homogenizer	→ Homogenizer.
Primary Electroviscous Effect	→ Electroviscous Effect.
Primary Emulsion	One way to make a multiple emulsion is to first create a reasonably stable first or "primary" emulsion using an appropriate surfactant and high shear mixing. The primary emulsion is then itself emulsified into more of the liquid that comprised the internal phase of the primary emulsion, this time with a different surfactant, and with low shear agitation. This produces a second, multiple emulsion. *See also* Reference [28].
Primary Minimum	In a plot of Gibbs energy of interaction versus separation distance, two minima can occur. The minimum occurring at the shortest distance of separation is referred to as the "primary minimum", and that occurring at larger separation distance is termed the "secondary minimum".
Primary Oil Recovery	The first phase of crude oil production, in which oil flows naturally to the wellbore. → Secondary Oil Recovery, → Enhanced Oil Recovery.
Primary Particle	The smallest identifiable particle in a sample of particles. → Secondary Particle.
Princen, Henricus (Mattheus) (1937–2004)	A Dutch colloid and interface scientist who worked in a range of industrial and academic settings and is best known for his contributions to understanding the rheology of foams and concentrated emulsions. Eponyms include the Princen model for the elastic behaviour of well-ordered two-dimensional systems. *See* Reference [198].
Probe Molecule	(Surfactant) Any species that is soluble in micelles and can be readily detected and measured. Example: Pyrene solubilized in micelles can be a reporter probe for its environment through fluorescence spectroscopy.
Profilometry	→ Scanning Probe Microscopy.

Promoter	In catalysis, any substance that interacts with a catalyst and causes an improvement in catalytic activity.
Protected Lyophobic Colloids	→ Sensitization.
Protection	The process in which a material adsorbs onto droplet surfaces and thereby makes an emulsion (or foam) less sensitive to aggregation and coalescence by any of a number of mechanisms, often via steric stabilization. → Sensitization.
Protective Colloid	A colloidal species that adsorbs onto and acts to "protect" the stability of another colloidal system. The term refers specifically to the protecting colloid and only indirectly to the protected colloid. Example: when a lyophilic colloid such as gelatin acts to protect another colloid in a dispersion by conferring steric stabilization. → Gold Number, → Protection.
Protein Colloid	Natural polymers, called "gelatin" or "animal glue", that are derived from collagen, a major protein in skin, tissues and bone. They are produced by partial hydrolysis of collagen and are amphoteric, containing both amine and acid groups. Gelatin and animal glue represent different degrees of purity; gelatin is more pure. They are examples of hydrophilic colloids. *See* Reference [136].
Protein Foam	(P) A fire-extinguishing foam based on hydrolyzed protein surfactants. → Fluoroprotein Foam, → Film Forming Fluoroprotein Foam, → Aqueous Film Forming Foam, → Alcohol Resisting Aqueous Film Forming Foam, → Fire Extinguishing Foam.
Protein Load	→ Surface Load.
PSD	(1) Particle size distribution. (2) Photon-Stimulated Desorption.
Pseudocolloid	A colloidal species or dispersion of colloidal species in which the colloidal unit has a an adsorbed component by which the colloid is detected or measured, as opposed to a "pure" colloidal species having no adsorbed material. Thus, pseudocolloids (also termed "Fremdkolloide" or "foreign colloid") have been distinguished from real or true colloids (also termed "Eigenkolloide" or "self-colloid"), although the use of all of these terms has been discouraged; *see* Reference [25], p. 72.
Pseudoemulsion Film	A fluid film of an aqueous phase (water) between air and oil phases. These are usually described as O/W/A or A/W/O, in which W represents the thin aqueous film. → Fluid Film.
Pseudophase Diagram	A phase diagram for a system in which more phases are present than are allowed to vary in the diagram. A pseudophase diagram is thus only one of several that are needed to completely describe a system.

Pseudoplastic	A non-Newtonian fluid whose viscosity decreases as the applied shear rate increases, a process that is also termed "shear thinning". Pseudoplastic behavior can occur in the absence of a yield stress and also after the yield stress in a system has been exceeded (i.e., once flow begins).
Pseudo-solution	$\rightarrow$ Colloidal, $\rightarrow$ Colloidal Dispersion.
Pulp	In mineral processing, a slurry of crushed ore dispersed in water.
Pure Colloid	$\rightarrow$ Pseudocolloid.
PVD	$\rightarrow$ Physical Vapour Deposition.
Pyruvic Acid Acetal	One of the kinds of polysaccharide structure constituting agar. Together with sulfated galactan, the combination is sometimes referred to as "charged agar", or "agaropectin". $\rightarrow$ Agar.

Qbit $\rightarrow$ Quantum Computer.

QD $\rightarrow$ Quantum Dot.

QELS   Quasi-Elastic Light Scattering. $\rightarrow$ Photon Correlation Spectros-
                        copy.

Quadrupole   The property of having the equivalent of two dipoles (electric or
                        magnetic), whose dipole moments have the same magnitude but
                        point in opposite directions and are separated from each other.

Quantum Bit $\rightarrow$ Quantum Computer.

Quantum Computer   A quantum computer uses quantum mechanical phenomena,
                        such as superposition, to perform data operations. In contrast to
                        conventional computer bits, which can have only two states, 0 or 1,
                        quantum bits (also termed qubits or qbits) can have three states, 0
                        or 1 or both at the same time. The latter state is called the
                        superposition state and allows multiple calculations to be con-
                        ducted simultaneously (the quantum parallelism phenomenon
                        suggested by Feynman). It has been suggested that this could be
                        achieved using laser light to excite appropriate atoms (to "write"
                        information) and monitoring photon emission (to "read" infor-
                        mation). Quantum computers have been made in which compu-
                        tations have been executed on a small number of qubits. It
                        is expected that large-scale quantum computers could func-
                        tion exponentially faster than conventional computers. *See*
                        Reference [36].

Quantum Dot   (QD) A nanoscale particle in which electrons are confined in all
                        three directions. Whereas bulk semiconductor crystals exhibit a
                        continuum of electron states, in a quantum well (QW) electrons

Dictionary of Nanotechnology, Colloid and Interface Science. Laurier L. Schramm
Copyright © 2008 WILEY-VCH Verlag GmbH & Co. KGaA, Weinheim
ISBN: 978-3-527-32203-9

are restricted to a foil that is a few nm thick, and in a quantum dot the electron states resolve into a set of discrete bands (or 'lines') that resemble the electron states of an atom. The quantization of electron energies in QDs allows them to be tuned to produce size-dependent optical and electrical properties. Quantum dots have been made from cadmium compounds like cadmium selenide, for example. They can also be made to have specific fluorescence emission colours, based on chemical composition and particle size. Quantum dots have also been termed mesoscopic atoms, or artificial atoms. Quantum dots are expected to lead to single-electron pump, biomolecule markers, QD laser, and nanoscale memory device applications. $\rightarrow$ Nanodot, $\rightarrow$ Quantum Computer, $\rightarrow$ Quantum Wire. *See* Reference [37].

Quantum Well	$\rightarrow$ Quantum Dot.
Quantum Wire	(QWR) A nanoscale wire or fibre in which electrons are confined in two dimensions, as opposed to three dimensions like in a quantum dot. $\rightarrow$ Quantum Dot.
Quasielastic Light Scattering	(QELS) $\rightarrow$ Light Scattering, $\rightarrow$ Photon Correlation Spectroscopy.
Quats	Quaternary ammonium compounds. These compounds are cationic surfactants if they contain a hydrocarbon chain of sufficient length. Example: cetyltrimethylammonium bromide, $CH_3(CH_2)_{15}N^+(CH_3)_3Br^-$. $\rightarrow$ Cationic Surfactant.
Qubit	$\rightarrow$ Quantum Computer.
QW	Quantum Well. $\rightarrow$ Quantum Dot.
QWR	$\rightarrow$ Quantum Wire.

R

Radiocolloid	Any colloidal species containing a radionuclide. Somewhat of a misnomer in that colloidal properties derive from the nature of atoms and molecules rather than from nuclei.
Radius of Gyration	(R_g) A measure of the effective size of a polymer chain in solution. The radius of gyration is the average distance from the center of gravity of a polymer molecule to a monomer segment (chain). The radius of gyration of a polymer in a poor solvent is therefore smaller than in a good solvent because the polymer molecule contracts and coils-up in the former case and uncoils and swells in the latter case. The radius of gyration for polymer molecules can be determined from light scattering results.
Raffinate	The phase that remains after specified solute(s) has been removed by extraction. Note that this meaning is distinct from the original meaning of raffinate as a refined product.
Rag Layer Emulsion	→ Interface Emulsion.
Rain	A rain cloud is an aerosol of liquid droplets. → Aerosol of Liquid Droplets, *see* Table 5.
Raman Effect	→ Raman Scattering.
Raman Scattering	When light is scattered from a solid some components of the scattered light have a shifted frequency due to energy gains or losses when atoms or molecules change vibrational or rotational motions. The Raman frequency is the difference between the frequency of scattered and incident light and gives the vibrational frequency of the bond responsible for the scattering.
Raman Spectroscopy	The use of Raman scattering to identify functional groups and bonding in a solid sample.

Dictionary of Nanotechnology, Colloid and Interface Science. Laurier L. Schramm
Copyright © 2008 WILEY-VCH Verlag GmbH & Co. KGaA, Weinheim
ISBN: 978-3-527-32203-9

Random Copolymer	A copolymer in which the constituent monomer molecules are randomly arranged in the polymer backbone. → Block Copolymer.
Rayleigh-Gans-Debye Scattering	A modified model of Rayleigh scattering. → Light Scattering.
Rayleigh Instability	→ Rayleigh-Taylor Instability.
Rayleigh, Lord	→ Strutt, John William
Rayleigh Ratio	In light scattering, the ratio of intensities of incident to scattered light at some specified distance.
Rayleigh Scattering	Light will be scattered (deflected) by local variations in refractive index caused by the presence of dispersed species and depending upon their size. The scattering of light by species whose size is much less than the wavelength of the incident light is referred to as "Rayleigh scattering", and it is termed "Mie scattering" if the species' size is comparable with that of the incident light. An example of Rayleigh scattering is that due to molecules in the atmosphere that scatter blue light from the sun's white-light illumination and cause the sky to appear blue while the sun appears orange-yellow. → Mie Scattering, → Light Scattering.
Rayleigh-Taylor Instability	The instability of an interface between two fluids of different densities caused by the acceleration of the less-dense fluid toward the more-dense fluid.
RBS	Rutherford Backscattering Spectroscopy. → Ion-Scattering Spectroscopy.
RCF	→ Relative Centrifugal Force.
Real Colloid	→ Pseudocolloid.
Receding Contact Angle	The dynamic contact angle that is measured when one phase is receding, or reducing its area of contact, along an interface while in contact with a third, immiscible phase. It is essential to state which interfaces are used to define the contact angle. → Contact Angle.
Reduced Adsorption	The relative Gibbs surface concentration of a component with respect to the total Gibbs surface concentration of all components. *See* Reference [4] for the defining equations.
Reduced Limiting Sedimentation Coefficient	→ Sedimentation.
Reduced Osmotic Pressure	→ Colloid Osmotic Pressure.
Reduced Sedimentation Coefficient	→ Sedimentation.

Reduced Viscosity	For solutions or colloidal dispersions, the specific increase in viscosity divided by the solute or dispersed-phase concentration, respectively ($\eta_{Red} = \eta_{SP}/C$). Also termed the "viscosity number". *See* Table 4.
Redwood Viscosity	A parameter intended to approximate fluid viscosity and measured by a specific kind of orifice viscometer. The measurement unit is Redwood seconds.
Reflected Light	Light that strikes a surface and is redirected back.
Reflection High-Energy Electron Diffraction	(RHEED) $\rightarrow$ High-Energy Electron Diffraction.
Refracted Light	Light that has changed direction by passing from one medium through another in which its wave velocity is different.
Regioselective Catalyst	A catalyst that increases the rate or yield of a reaction when the reaction occurs at specific sites on the catalyst. Also termed "regiospecific catalyst".
Relative Adsorption	The relative Gibbs surface concentration of a component with respect to that of another specified component. *See* Reference [4] for the defining equations.
Relative Centrifugal Force	(RCF) When a centrifuge is used to enhance sedimentation or creaming, the centrifugal force is equal to mass times the square of the angular velocity times the distance of the dispersed species from the axis of rotation. The square of the angular velocity times the distance of the dispersed species from the axis of rotation, when divided by the gravitational constant, g, yields the relative centrifugal force or RCF. RCF is not strictly a force but rather the proportionality constant. It is substituted for g in Stokes' law to yield an expression for centrifuges and is used to compare the relative sedimentation forces achievable in different centrifuges. Because RCF is expressed in multiples of g, it is also termed "g-force" or simply "gs".
Relative Micellar Mass	$\rightarrow$ Relative Molecular Mass of Micelles.
Relative Micellar Weight	$\rightarrow$ Relative Molecular Mass of Micelles.
Relative Molecular Mass	The mass of 1 mole of species (actually it is the mass of 1 mole of species divided by the mass of 1/12 mole of ^{12}C). Relative molecular mass has replaced the older term "molecular weight". $\rightarrow$ Number-Average Quantities, $\rightarrow$ Mass-Average Quantities.
Relative Molecular Mass of Micelles	The mass of 1 mole of micelles (actually it is the mass of 1 mole of micelles divided by the mass of 1/12 mole of ^{12}C). Other terms that mean the same thing are "relative micellar mass" and "relative micellar weight".

Relative Permittivity	$\rightarrow$ Permittivity, *see* Table 9.
Relative Viscosity	In solutions and colloidal dispersions, the viscosity of the solution or dispersion divided by the viscosity of the solvent or continuous phase, respectively ($\eta_{Rel} = \eta/\eta_o$). Also termed the "viscosity ratio". *See* Table 4.
Relative Viscosity Increment	$\rightarrow$ Specific Increase in Viscosity.
Relaxation Effect	$\rightarrow$ Electrophoretic Relaxation.
Relaxation Time	The time required for the value of a changing property to be reduced to $1/e$ of its initial value.
Repeptization	Peptization, usually by dilution, of a once-stable dispersion that was aggregated (coagulated or flocculated) by the addition of electrolyte.
Replica	A metal film duplicate of a sample used in scanning electron microscopy. For example, an emulsion sample can be fast-frozen in a cryogen, fractured to reveal interior structure, then coated with a metal film to preserve the structure. $\rightarrow$ Freeze-Fracture Method.
Replicator	(Nanotechnology) A conceptual device ($\rightarrow$ Drexler, (Kim) Eric) that would be a specialized kind of assembler, capable of making exact copies of itself. Nanotechnology replicators would work in analogous fashion to the self-replication conducted by DNA in living cells. Also termed nanoreplicator.
Reporter Probe	$\rightarrow$ Probe Molecule.
Repulsive Potential Energy	$\rightarrow$ Gibbs Energy of Repulsion.
Residual DNAPL	The dense nonaqueous-phase liquid (DNAPL) left behind in a soil that has been treated by a soil flushing process intended to remove the DNAPL. Bothe the flushing process and the residual phase are analogous to an *in situ* enhanced oil recovery process and the left-behind residual crude oil. $\rightarrow$ Enhanced Oil Recovery, $\rightarrow$ Nonaqueous-Phase Liquid.
Resin	In surface coatings or paint, $\rightarrow$ Binder.
Resistazone Counter	$\rightarrow$ Sensing-Zone Technique.
Resolution	In emulsion treatment, resolution refers to emulsion breaking and the separation of the oleic and aqueous phases. Example: The breaking and separation of oilfield produced W/O emulsions.
Retardation Effect	$\rightarrow$ Electrophoretic Retardation.
Retarded van der Waals Constant	$\rightarrow$ van der Waals Constant.
Retentate	$\rightarrow$ Dialysis, $\rightarrow$ Electrodialysis, $\rightarrow$ Filtration.

Reverse Emulsion	A petroleum industry term used to denote an oil-in-water emulsion (most wellhead emulsions are W/O). Reverse emulsion has the opposite meaning of the term "invert emulsion". → Invert Emulsion.
Reverse Flotation	In contrast to conventional flotation, in which the desirable mineral is directly floated and collected from the produced froth, in reverse flotation the undesirable minerals are preferentially floated and removed leaving behind a slurry that has been concentrated in the desirable mineral. This method has been used for the removal of salt from potash. Also termed indirect flotation.
Reverse Micelles	Synonym for the dispersed phase in a water-in-oil type microemulsion. Here the surfactant heads, or polar groups, associate closely to minimize interaction with the oil phase. This close association can happen when they orient themselves inside water droplets, and it also allows the surfactant tails, or hydrocarbon groups, to stabilize the water droplets by orienting toward or into the oil.
Reverse Osmosis	(RO) A separation process like ultrafiltration except applied to sub-colloidal particles. A solution is separated from pure liquid by a semipermeable membrane, and a pressure applied across the membrane drives the separation. Example: desalination. → Dialysis, → Filtration.
Reversible Sol–Gel Transformation	→ Thixotropic.
Reynolds Number	A dimensionless quantity used in the modelling of systems in which viscosity plays a role in determining the velocities, or flow profile, of a fluid. Its value is used to distinguish laminar from turbulent flow regimes.
Reynolds, Osborne (1842–1912)	An Irish-born U.K. scientist and mathematician known for his work in fluids, hydraulics and heat transfer. On the applied side he developed several kinds of pumps, brakes and steam engines. He is remembered in rheology through the "Reynolds number" and the concepts of laminar and turbulent flow.
R_g	→ Radius of Gyration.
RHEED	Reflection High-Energy Electron Diffraction. → High-Energy Electron Diffraction.
Rheology	The science of deformation and flow of all forms of matter. The term rheology was coined by E.C. Bingham and M. Reiner in 1929 [199]. Rheological descriptions usually refer to the property of viscosity and departures from Newton's law of viscosity. → Rheometer.

Rheomalaxis	A special case of time-dependent rheological behavior in which shear-rate changes cause irreversible changes in viscosity. The change can be negative, as when structural linkages are broken, or positive, as when structural elements become entangled (like work-hardening).
Rheometer	Any instrument designed for the measurement of non-Newtonian and Newtonian viscosities. The principal class of rheometer consists of the rotational instruments in which shear stresses are measured while a test fluid is sheared between rotating cylinders, plates, or cones. Examples of rotational rheometers: concentric cylinder, cone-cone, cone-plate, double-cone-plate, plate-plate, and disc. *See* Reference [21].
Rheopexy	Dilatant flow that is time-dependent. At constant applied shear rate, viscosity increases, and in a flow curve, hysteresis occurs (opposite to the thixotropic case). The term rheopexy was coined by Freundlich and Juliusberger in 1935 [200]. Also termed antithixotropy. Rheopexy is termed "work hardening" if the original viscosity is not recovered after the shear is discontinued.
Richardson Equation	(1) An empirical equation for emulsion viscosity. *See* Table 11. (2) An expression giving the critical velocity needed for a liquid jet to break apart into droplets and thus form an emulsion. → Ohnesorge Equation.
Rigid Film	→ Fluid Film.
Ring-Roller Mill	A machine for the comminution, or size reduction, of minerals. Such machines crush the input material between a stationary ring and vertical rollers revolving inside the ring. Particle sizes of as low as about $30\,\mu m$ can be produced.
RO	→ Reverse Osmosis.
Robertson-Stiff Equation	A rheological flow model in which shear stress is given with two constants and the power law model. This model has been applied to practical dispersions, including suspensions, emulsions, and foams. For this and other models *see* References [25, 28]. → Herschel-Bulkley Equation.
Rock Flour	Finely ground particles of quartz and feldspar produced by the crushing and grinding action of glaciers on underlying rocks, or when freeze thaw cycles break up rocks. These particles form colloidal suspensions in natural waters such as rivers and lakes and may lend green or turquoise colours to such lakes.
Roll Crusher	A machine for the comminution, or size reduction, of mineral lumps or stones. Such machines crush the input material between a plate and revolving roller or between more than one roller.

Roll Mill	A device for imparting shear to a dispersion for the purpose of reducing the particle size of the dispersed material. Somewhat similar to a roll crusher. Examples of dispersions processed over a roll mill are pigment grinds (pigments dispersed in a fluid such as castor oil) and soap formulations (where the dispersed material includes fragrance oil droplets and pigments).
Ross Foam	Foam produced from a binary or ternary solution under conditions at which its temperature and composition approach (but do not reach) the point of phase separation into separate immiscible liquid phases.
Ross–Miles Test	A method for assessing foam stability in which one measures the rate of collapse of a (static) column of foam that has been generated by allowing a certain quantity of foaming solution to fall a specified distance into a separate volume of the same solution contained in a vessel. This technique is ASTM method D1173-53. $\rightarrow$ Static Foam Test, $\rightarrow$ Bartsch Test.
Rotational Rheometer	$\rightarrow$ Rheometer.
Roughing Flotation	A first-pass flotation of desired minerals that are well liberated, well conditioned, and are readily floated. Roughing flotation is followed by another flotation step, intended to separate other component(s) from the ore. $\rightarrow$ Carrier Flotation, $\rightarrow$ Emulsion Flotation, $\rightarrow$ Floc Flotation, $\rightarrow$ Scalping Flotation, $\rightarrow$ Scavenging Flotation, $\rightarrow$ Froth Flotation.
Roughness Factor	The factor by which the surface area of a nonporous solid is greater than that calculated from the macroscopic dimensions of the surface.
Rubber Latex	Natural rubber latex is rubber tree sap. It is a colloidal system of cis-1,4-polyisoprene particles dispersed in an aqueous serum, and has a cloudy white appearance. In freshly tapped latex the particles are covered by a layer of phospholipid and protein molecules that are negatively charged and provide the colloidal stability. $\rightarrow$ Latex.
Rule of Schulze and Hardy	$\rightarrow$ Schulze–Hardy Rule.
Rupture	$\rightarrow$ Fluid Film.
Rutherford Backscattering Spectroscopy	(RBS) $\rightarrow$ Ion-Scattering Spectroscopy.

S

SAD	→ Surfactant Affinity Difference.
SAE Viscosity	A Society of Automotive Engineers (SAE) motor oil viscosity classification. SAE summer grades are defined by viscosity ranges. For example, SAE 30 oil will have a minimum low-shear rate, kinematic viscosity of 9.3 cSt (at 100 °C), and a maximum of less than 12.5 cSt (at 100 °C), and also will have a minimum high-shear rate absolute viscosity of 2.9 cP (at 150 °C). SAE winter grades are defined by, in addition, maximum viscosities at specified low temperatures, and carry a W suffix. Multigrade oils meet combined winter and summer specifications and have combined designations such as SAE 10 W-30.
Salinity Requirement	→ Optimum Salinity.
Salt Curve	A graphical representation of the viscosity of a system versus salt concentration. This curve can be an important characteristic of formulated systems in which viscosity control is necessary, such as in shampoo formulas.
Salting In	(1) Solutions: When the addition of electrolyte to a solution causes an increase in the solubility of a solute. → Salting Out.
	(2) Surfactants: When the addition of electrolyte to a solution of nonionic surfactant causes the critical micelle concentration to increase. Also, addition of electrolyte to an ionic surfactant solution in a multiphase system can drive surfactant from the oil phase into the aqueous phase. → Salting Out.
Salting Out	(1) Solutions: When the addition of electrolyte to a solution causes a decrease in the solubility of a specified solute. → Salting In.
	(2) Surfactants: When the addition of electrolyte to a solution of nonionic surfactant causes the critical micelle concentration to decrease. Also, addition of electrolyte to an ionic surfactant solution

Dictionary of Nanotechnology, Colloid and Interface Science. Laurier L. Schramm
Copyright © 2008 WILEY-VCH Verlag GmbH & Co. KGaA, Weinheim
ISBN: 978-3-527-32203-9

	in a multiphase system can drive surfactant from the aqueous phase into the oil phase. → Salting In. (3) Emulsions: The process of demulsification by the addition of electrolyte.
Sand	A term used to distinguish particles having different sizes in the range between about 50–63 µm and about 2000 µm, and with several subcategories, all depending on the operational scale adopted. *See* Table 12.
SANS	→ Small Angle Neutron Scattering
Saponification	The reaction of a fat or a fatty acid with a base to produce soap.
Saponification Number	An indication of the amount of fatty saponifiable material in a sample, usually in an oil. It is given as the number of milligrams of potassium hydroxide (KOH) that combine with one gram of sample (oil) under specified test conditions, such as by test method ASTM D 94. Caution must be used in interpreting test results if the samples contain other substances that react with KOH, such as sulfur compounds or halogens, as these will increase the apparent saponification number.
Saybolt Furol Seconds	(SFS). → Saybolt Furol Viscosity.
Saybolt Furol Viscosity	(SFV) A parameter intended to approximate fluid viscosity and measured by a specific kind of orifice (tube) viscometer. The viscometer has a larger diameter tube than does a Saybolt Universal viscometer, and is used for very viscous fluids, such as heavy oils. → Saybolt Universal Viscosity.
Saybolt Universal Seconds	(SUS). → Saybolt Universal Viscosity.
Saybolt Universal Viscosity	(SUV) A parameter intended to approximate fluid viscosity and measured by a specific kind of orifice (tube) viscometer. Used for low density petroleum products. → Saybolt Furol Viscosity.
Scalping Flotation	A first flotation process step, that is used to remove a minor, hydrophobic mineral from the ore. This mineral could be one that is valuable, such as MoS_2 from Cu-Mo ore, or one that is unwanted, such as talc. Scavenging flotation is followed by another flotation step, intended to separate a major component from the ore. → Carrier Flotation, → Emulsion Flotation, → Floc Flotation, → Roughing Flotation, → Scavenging Flotation, → Froth Flotation.
Scanning Electron Microscopy	(SEM) → Electron Microscopy.
Scanning Force Microscopy	→ Friction Force Microscopy.

Scanning High-Energy Electron Diffraction	(SHEED) → High-Energy Electron Diffraction.

Scanning Near-Field Optical Microscopy

(SNOM) → Near-Field Scanning Optical Microscopy.

Scanning Probe Microscopy

(SPM) A family of microscopies that use the interaction of a probe with a surface in order to create an image. The probe tip is held at the end of a cantilever, and the deflection of the cantilever is measured by a device called an optical lever. In scanning tunnelling microscopy (STM) the metal probe, sharpened to a few tenths of a nanometer at the tip, is brought very close to a surface so that a current is generated resulting from overlapping wave functions (electron tunneling). The current varies with separation distance, and thus surface morphology can be determined as the tip is caused to move over the surface; the technique is referred to as "profilometry". In an atomic force microscope (AFM), force rather than current is measured. In dynamic force microscopy (DFM) the tip is made to oscillate perpendicular to the sample's surface (hence dynamic, atomic force microscopy). Friction force microscopy (FFM) is atomic force microscopy in which forces are measured along the scanning direction and used in micro- and nanoscale studies of friction and lubrication. Magnetic force microscopy (MFM) uses a magnetic tip to probe the magnetic field above a sample surface. → Near-Field Scanning Optical Microscopy for an optical probe version of SPM. → Nanomanipulator, *see* Table 7.

Scanning Probe Surface Patterning

Any of a number of techniques that employ an atomic force microscope (AFM) to deliver energy to a surface in a way that results in a pattern. Examples include anodic oxidation, in which an AFM tip is used to selectively oxidize a surface to make a pattern, nanografting, in which the tip is used to remove a monolayer resist to make a pattern, nanoelectrochemical patterning, in which the tip is used to electrochemically convert the end groups of molecules in a monolayer into a form that can be chemically modified to make a pattern, and thermomechanical writing, in which a heated tip is used to thermally modify a polymer layer to make a pattern. These patterning techniques all use energy to make the patterns and are surface-destructive. For surface-constructive patterning techniques → Nanolithography. *See* Reference [90].

Scanning Transmission Electron Microscopy

(STEM) → Electron Microscopy.

Scanning Tunneling Microscopy

(STM) → Scanning Probe Microscopy. *See also* Table 7.

Scattering Angle	→ Light Scattering.
Scattering Plane	→ Light Scattering.
Scavenging Flotation	A flotation separation process, in which particles or droplets become attached to gas bubbles that are injected (sparged) into the flotation medium. Scavenging flotation is a subsequent flotation step that is applied to the tailings from the roughing stage. This step is intended to float those minerals that require longer treatment time or otherwise different flotation operating conditions from the roughing stage. Sometimes termed "induced gas flotation" although this term is less specific. Example: the scavenging flotation of bitumen from oil sand process tailings. → Carrier Flotation, → Emulsion Flotation, → Floc Flotation, → Roughing Flotation, → Scalping Flotation, → Froth Flotation.
Schiller Layers	The layers of particles that can be formed during sedimentation such that the distances between layers are on the order of the wavelength of light, leading to iridescent, or Schiller, layers.
Schlieren Optics	An optical arrangement designed to allow detection of density gradients occurring in fluid flow. Typically a narrow beam of light is collimated by one lens and focused on a knife-edge by a second lens. A density gradient in a fluid, between the lenses, causes a diffraction pattern to appear beyond the knife-edge.
Schulze, Hans (Joachim) (1938–2003)	A German colloid and interface scientist known for his work on fundamental and applied aspects of mineral processing and flotation, much of which is included in his well-known book "Physico-Chemical Elementary Processes in Flotation." *See* References [201, 202].
Schulze–Hardy Rule	An empirical rule summarizing the general tendency of the critical coagulation concentration (CCC) of a suspension, an emulsion, or other dispersion to vary inversely with about the sixth power of the counterion charge number of added electrolyte. Also termed the "sixth-power law". *See* References [144, 203, 204]. → Critical Coagulation Concentration.
Scratch Hardness	→ Hardness.
Screening Length	→ Debye Length.
Sculptured Nematic Thin Film	(SNTF) → Sculptured Thin Film.
Sculptured Thin Film	(STF) Physical vapour deposition (PVD) is used to make a nanosize film by depositing material on a substrate, in successive rows. Columnar thin films (CTFs) are a kind of STF in which PVD is used to create a thin-film layer comprising angled columns anchored to the substrate. Other kinds of STFs are made by changing

the column direction in various ways, during the deposition process. Among the different morphologies that can be produced are sculptured nematic thin films (SNTFs) and thin-film helicoidal bianisotropic media (TFHBM). *See* Reference [37].

SEAR	→ Surfactant Enhanced Aquifer Remediation.
Seaweed Colloids	A class of hydrophilic colloids (hydrocolloids) derived from various seaweeds. This class includes agar, algin, furcellaran, and carrageenan.
Secondary Electrons	In electron spectroscopy, electrons leaving a surface that are produced by energy transfer from the incident beam. Also applied to electrons having energies that are less than 50 eV.
Secondary Electroviscous Effect	→ Electroviscous Effect.
Secondary Ion Mass Spectroscopy	(SIMS) A technique for studying surface composition in which a beam of very high-energy ions strikes a surface and causes ions from the surface to be ejected. The masses of the ejected ions are determined in a mass spectrometer; hence surface compositions can be determined. *See* Reference [57] for specific terms in SIMS.
Secondary Minimum	→ Primary Minimum.
Secondary Oil Recovery	The second phase of crude oil production, in which water or an immiscible gas are injected to restore production from a depleted reservoir. → Primary Oil Recovery, → Enhanced Oil Recovery.
Secondary Particle	A particle formed by the aggregation of primary particles.
Second Harmonic Generation	→ Optical Second Harmonic Generation.
Sediment	The process of sedimentation in a dilute dispersion generally produces a discernable, more concentrated dispersion, which is termed the "sediment" and has a volume termed the "sediment volume".
Sedimentation	The settling of suspended particles or droplets due to gravity or an applied centrifugal field. The rate of this settling is the sedimentation rate (or velocity). The sedimentation rate divided by acceleration is termed the "sedimentation coefficient". The sedimentation coefficient extrapolated to zero concentration of sedimenting species is termed the "limiting sedimentation coefficient". The sedimentation coefficient reduced to standard temperature and solvent is termed the "reduced sedimentation coefficient". If extrapolated to zero concentration of sedimenting species it is termed the "reduced limiting sedimentation coefficient". Negative sedimentation is also called "flotation". Flotation in which droplets

rise upwards is also called "creaming". Flotation in which particulate matter becomes attached to gas bubbles is also referred to as "froth flotation". → Creaming, → Froth Flotation, → Subsidence.

Sedimentation Coefficient	→ Sedimentation.
Sedimentation Equilibrium	The state of a colloidal system in which sedimentation and diffusion are in equilibrium.
Sedimentation Field Strength	→ Sedimentation Potential.
Sedimentation Potential	The potential difference at zero current caused by the sedimentation of dispersed species. This mechanism of potential difference generation is known as the "Dorn effect"; accordingly, the sedimentation potential is sometimes referred to as the "Dorn potential". The sedimentation can occur under gravitational or centrifugal fields. The potential difference per unit length in a sedimentation potential cell is the sedimentation field strength. This method can be used to determine zeta potential. *See also* Table 16.
Sediment Volume	→ Sediment.
Selective Flocculation	A mineral processing technique for the separation of valuable components from other, less valuable, components (gangue). A polymer is added that will selectively adsorb onto one or the other of the components and thereby induce flocculation (aggregation) of that component. The flocs are then easily separated from other components, usually by sedimentation.
Self-Assembled Monolayers	(SAMs) Monolayers formed on a surface by a self-assembly process driven, not by hydrophobic interactions, but by a specific and stronger interaction. Example: SAMs have been formed between alkanethiols and a gold surface. Here, the thiol group bonds to the gold. Changing the nature of the hydrocarbon tail group permits changing the nature and properties of the surface monolayer. Beyond colloidal scale surface properties and applications, SAMs are being applied in molecular electronic devices and nanolithography.
Self-Assembling Colloid	→ Association Colloid.
Self-Assembly	The spontaneous assembly of molecules into organized structures such as micelles or adsorption layers. Self-assembly can be driven by chemical or physical bonding. Also termed molecular self-assembly. → Solution Phase Templating, → Nanolithography.
Self-Beat Light Scattering	→ Photon Correlation Spectroscopy.
Self-Colloid	→ Pseudocolloid.

Self-Diffusion Coefficient	→ Diffusion Coefficient.
Self-Organizing	→ Self-Assembly.
SEM	Scanning electron microscopy. → Electron Microscopy.
Semi-Colloid	→ Lyophilic Colloid.
Semiconducting Nanowire	→ Nanowire.
Semipermeable Membrane	A membrane that is permeable to some species and not to others according to species size and electric charge. → Colloid Osmotic Pressure.
Sensing-Zone Technique	A general term used to refer to any of the particle- or droplet-sizing techniques that rely on (usually) conductivity or capacitance changes in sample introduced between charged electrodes. Also termed "resistazone counter". An example is the Coulter counter. The term "sensing zone technique" is also used with reference to similar techniques that use light absorption or scattering instead of electrical properties, → Photozone Counter.
Sensitization	The process in which small amounts of added hydrophilic colloidal material make a hydrophobic colloid more sensitive to coagulation by electrolyte. Example: the addition of polyelectrolyte to an oil-in-water emulsion to promote demulsification by salting out. Higher additions of the same material usually make the emulsion less sensitive to coagulation, and this is termed "protective action" or "protection". The protected, colloidally stable dispersions that result in the latter case are termed "protected lyophobic colloids".
Separator	In the petroleum industry, a vessel designed to separate the oil phase in a petroleum fluid from some or all of the other three constituent phases: gas, solids, and water. Free-water knockouts fall under this category, but so do separators capable of breaking and removing water and solids from emulsions. The latter range from gravity to impingement (coalescence) to centrifugal separators.
Septum	In general, any dividing wall between two cavities. Example: the thin liquid films (lamellae) between bubbles in a foam.
Sessile Bubble Method	→ Sessile Drop Method.
Sessile Drop Method	A method for determining surface or interfacial tension based on measuring the shape of a droplet at rest on the surface of a solid substrate (in liquid-liquid systems the droplet can alternatively rest upside down, that is, underneath a solid substrate). This technique can also be used to determine the contact angle and contact diameter of the droplet against the solid.

SET	Single-Electron Transistor. → Single-Electron Device.
Settling Radius	→ Equivalent Spherical Diameter.
SEXAFS	Surface Extended X-Ray Absorption Fine Structure Spectroscopy. → Extended X-Ray Absorption Fine Structure Spectroscopy.
SFA	→ Surface Force Apparatus.
SFS	Saybolt Furol Seconds. → Saybolt Furol Viscosity.
SFV	→ Saybolt Furol Viscosity.
Shear	The rate of deformation of a fluid when subjected to a mechanical shearing stress. In simple fluid shear, successive layers of fluid move relative to each other such that the displacement of any one layer is proportional to its distance from a reference layer. The relative displacement of any two layers divided by their distance of separation from each other is termed the "shear" or the "shear strain". The rate of change with time of the shear is termed the "shear rate" or the "strain rate". Some shear rates appropriate to various processes are given in Table 19.
Shear Adhesion	→ Peel Test.
Shear Plane	Any species undergoing electrokinetic motion moves with a certain immobile part of the electric double layer that is commonly assumed to be distinguished from the mobile part by a sharp plane, the shear plane. The shear plane is also termed the "surface of shear". The zeta potential is the potential at the shear plane. → Zeta Potential.
Shear Rate	→ Shear.
Shear Stress	A certain applied force per unit area is needed to produce deformation in a fluid. For a plane area around some point in the fluid and in the limit of decreasing area the component of deforming force per unit area that acts parallel to the plane is the shear stress. Some shear rates appropriate to various processes are given in Table 19.
Shear Thickening	When the viscosity of a non-Newtonian fluid increases as the applied shear rate increases. → Dilatant.
Shear Thinning	When the viscosity of a non-Newtonian fluid decreases as the applied shear rate increases. → Pseudoplastic.
SHEED	Scanning High-Energy Electron Diffraction. → High-Energy Electron Diffraction.
Shell Viscosity	A parameter intended to approximate fluid viscosity and measured by a specific kind of orifice viscometer. The measurement unit is Shell seconds.

SHG	Second Harmonic Generation. $\rightarrow$ Optical Second Harmonic Generation.
Shortening	This term originally referred to the lard used to make bread and pies. The term shortening more commonly now refers to any commercial fat or oil used to tenderize baked foods by preventing the cohesion of wheat gluten strands during mixing. In this way the gluten strands can be considered to be "shortened" compared with their state if shortening had not been employed. $\rightarrow$ Margarine.
Short-Range Interaction Forces	Short-range (usually) repulsive forces that act in addition to van der Waals and electrostatic interaction forces, and at separation distances of the order of 2 to 3 nm. These forces have been called "solvation" or "hydration" forces and have been attributed to perturbations of the orientation of solvent molecules at surfaces, which hinder the approach of solvated species at short distances. Overlapping of the perturbed water molecules can lead to forces that are either attractive (hydrophobic forces) or repulsive (hydration forces) in nature [157, 205]. Effects such as molecular protrusion and conformation, can also contribute to short-range interaction forces.
SHS	$\rightarrow$ Super-Hydrophobic Surface.
Sibree Equation	An empirical equation for the viscosity of emulsions. *See* Table 11.
Sieve	$\rightarrow$ Particle Size Classification.
Silica Nanobottle	$\rightarrow$ Nanobottle.
Silica Nanofoam	$\rightarrow$ Nanofoam.
Silicone Oil	Any of a variety of silicon-containing polymer solutions. An example is a linear poly(dimethylsiloxane): $HO[(CH_3)_2SiO]_nH$.
Silt	A term used to distinguish particles having sizes of greater than about 2–4 μm and less than about 50–63 μm, depending on the operational scale adopted. *See* Table 12.
Silver Film	$\rightarrow$ Black Film.
SIMS	$\rightarrow$ Secondary Ion Mass Spectroscopy.
Singer Equation	An equation for the surface pressure exerted by a polymer monolayer in terms of the area per molecule and other properties of the polymer.
Single-Electron Device	Electronic devices through which electrons pass one at a time. Early single electron transistors (SET) have been made using thin films across which electrons "tunnel" (as in the quantum mechanical action tunnelling), and also by bending carbon nanotubes in

	multiple places in order to restrict electron flow. → Molecular Circuitry.
Single-Electron Transistor	(SET) → Single-Electron Device.
Single-Walled Carbon Nanotube	(SWCNT) → Carbon Nanotube.
Single-Walled Nanotubes	→ Nanotube.
Sink-Float Method	→ Hydrophobic Index.
Sintering	The coalescence or merging of two or more solid particles into a single particle.
Slime Coating	(1) A mineral flotation technique in which the flotation of coarse particles is decreased or prevented by coating their surfaces with fine hydrophilic particles (slimes). Slime coating is essentially the opposite of carrier flotation. Example: the slime coating of fine fluorite particles onto galena particles to prevent their flotation. → Adagulation. (2) In biology, slime coating is a term for the encapsulating slime secretion from a collection of microorganisms attached on a surface. Also known as a protective biofilm. Example: the slime on stones in a river.
Slimes Fraction	A suspension of small size fraction mineral particles. The size cut-off should be specified, such as <10 μm diameter (or −200 mesh). In mineral processing slimes represent a fine tailings by-product of the wet crushing and slurrying of minerals.
Slip-Casting	In ceramics, the process in which a slurry of dispersed particles is poured into a porous mould and allowed to stand. Water diffuses by capillary flow from the suspension into the mould causing the suspension to gel, and then to dry. The particles can then be sintered to form a final product.
Slipping Plane	The boundary layer that separates a stationary from a mobile phase when a fluid moves across a surface.
SLLS	→ Surface Laser Light Scattering.
Slow Sand Filtration	→ Filtration.
Slugging Compound	→ Knockout Drops.
Slurry Quality	In suspensions, the concentration of solid particles. In foams containing solids, the volume fraction of gas plus solid in the foam. → Foam Quality.
Small Angle Neutron Scattering	(SANS) A technique similar to light scattering using small angles compared with the incident beam. In this case a neutron beam is employed, to yield a shorter wavelength than visible light, in order

| | to obtain information about different regions on small particles, including the thickness of adsorption layers. *See also* Table 7. |

Smart Colloids
Colloidal dispersions that are able to respond in desirable ways to induced changes in physical conditions, such as temperature, illumination, solution properties, or an applied magnetic or electric field [206]. Smart colloids are a sub-set of smart materials. → Electroactive Solids, → Electro-Rheological Colloids, → Magneto-Rheological Colloids, → Thermo-Shrinking Polymers. *See* References [207, 208].

Smart Materials
Materials that are able to respond in desirable ways changes in physical (environmental) conditions.

Smectic State
→ Mesomorphic Phase.

Smog
Aerosol air pollution, including dispersed liquids and solids having diameters less than about $2\,\mu m$.

Smoke
A special kind of aerosol that results from a thermal process such as combustion or thermal decomposition. A smoke aerosol can be of solid particles or of liquid droplets. The term fume is sometimes used to denote the simultaneous presence of a gas. → Aerosol, and *see* Tables 5 and 6.

Smoluchowski, Marian (1872–1917)
An Austrian-born Polish physicist known for the theory of density fluctuations and demonstrations in which phenomena such as light scattering and the Brownian movement of colloidal particles supported the atomic-molecular nature of matter. He also worked in diffusion, coagulation, and electrokinetics. Eponyms include the "Smoluchowski equation" (in electrophoresis and in diffusion). *See* Reference [209].

Smoluchowski Equation
(1) Electrophoresis: A relation expressing the proportionality between electrophoretic mobility and zeta potential for the limiting case of a species that can be considered to be large and have a thin electric double layer. Also termed Helmholtz-Smoluchowski Equation. → Hückel Equation, → Henry Equation, → Electrophoresis.
(2) Diffusion: A relation expressing the proportionality between the rate of diffusional encounters between species and their diffusion coefficient, in which the constant of proportionality includes the radius and number concentration of the species. There are also derived Smoluchowski equations for specific processes, such as for rate of aggregation of particles.

Snap-Off
A mechanism for foam lamella generation in porous media. When gas enters and passes through a constriction in a pore, a capillary pressure gradient is created. The capillary pressure causes liquid to flow toward the region of the constriction, where it accumulates

and can cause the gas to pinch off or snap off and thereby create a new gas bubble separated from the original gas by a liquid lamella. → Lamella Division, → Lamella Leave-Behind.

SNOM — Scanning Near-Field Optical Microscopy. → Near-Field Scanning Optical Microscopy.

SNTF — Sculptured Nematic Thin Film. → Sculptured Thin Film.

Soap — A surface-active fatty acid salt containing at least eight carbon atoms. The term is no longer restricted to fatty acid salts originating from natural fats and oils. → Surfactant, *see* Table 14.

Soap Curd — A mixture of soap crystals in a saturated solution in which the soap crystals produce a gel-like consistency. The soap crystals in this case are referred to as "curd-fibres". Soap curd is not a mesomorphic (liquid-crystal) phase.

Soap Film — A thin film of water in air that is stabilized by surfactant. The term is used even though the film is not a film of soap and even where the surfactant is not a soap. → Fluid Film.

Soap Sticks — Solid, surfactant-containing sticks used to treat petroleum industry gas wells that have become laden with water at the bottom of the well. The sticks are dropped down into a well to where they dissolve and release the surfactant. The surfactant causes a reduction in surface tension that in turn reduces the gas velocity needed to lift the water out of the well. In some cases the surfactant is also used to generate foam to make it easier to lift water from the well. *See also* Reference [210], → Turner Velocity.

Soft X-Ray Appearance Potential Spectroscopy — (SXAPS) → Appearance Potential Spectroscopy.

Soft X-Ray Emission Spectroscopy — (SXES) → X-Ray Emission Spectroscopy.

Soil — Naturally occurring unconsolidated material, whether mineral or organic, that is on the earth's surface and is capable of supporting plant growth.

Sol — A colloidal dispersion. In some usage the term "sol" is meant to distinguish dispersions in which the dispersed-phase species are of very small size so that the dispersion appears transparent.

Solid Aerosol — → Aerosol of Solid Particles.

Solid Emulsion — A colloidal dispersion of a liquid in a solid. Example: opals, which contain dispersed water droplets. *See* References [7, 211].

Solid Foam — A colloidal dispersion of a gas in a solid. Examples: polystyrene foam, polyurethane foam. → Closed-Cell Foam, → Meerschaum.

Sol-Gel Processing	The sol-gel process refers to the transition of a dispersion of colloidal particles into gel state (a branched, continuous network). The gels are then typically dried out using, for example, supercritical drying, to produce aerogels that preserve most of the original open structure, and therefore low to ultralow density. Example: sol-gel processing can be used to make porous glass. *See* Reference [212]. → Aerogel, → Xerogel.
Solidification Front Method	A method for determining solid surface or interfacial tension. When a small particle, initially embedded in a liquid phase, is slowly approached by the solid-liquid interface of the slowly solidifying (freezing) liquid, it will remain embedded or become rejected by the front depending upon the free energy of adhesion, which in turn depends upon the solid-liquid interfacial tension. Also termed the "freezing front method". *See also* the review by Spelt [213].
Solid Suspension	A colloidal dispersion of a solid in another solid. Example: ruby-stained glass, a dispersion of gold particles in glass.
Solloids	Surface colloids. Colloidal-sized aggregates of surfactant and/or polymer species adsorbed on a surface. Used as a more general term than "admicelles" or "hemi-micelles". *See* Reference [214].
Solubility Parameter	A means of estimating solubility. It is the square root of the ratio of energy of vapourization to molar volume. This ratio is also known as the cohesive energy density. Pairs of substances having very similar solubility parameters tend to be mutually soluble.
Solubilizate	The solute whose solubility is increased in the process of solubilization.
Solubilization	The process by which the solubility of a solute is increased by the presence of another solute. Micellar solubilization refers to the incorporation of a solute (solubilizate) into or on micelles of another solute to thereby increase the solubility of the first solute.
Solubilizing Agent	Any product that can be used to aid in the solubilization of a species. Examples: solubilizing agents for dyestuffs or pigments. Often a surfactant, such as a fatty acid derivative. → Dispersant.
Solution Phase Templating	A method for producing structures in solution that uses molecular self-assembly together with a template.
Solution Precipitation	The mixing together of two or more solutions causing precipitation.
Solvation Forces	→ Short-Range Interaction Forces.
Solvent-Motivated Sorption	Sorption that occurs as a result of the hydrophobicity of a compound. Accumulation of the compound at the interface or in the

other phase is not due to its affinity for that phase so much as to its disaffinity for the initial phase. Such sorption occurs for organic contaminants in the environment. This kind of sorption can often be related to the octanol-water partitioning coefficient. → Hydrophobic Organic Contaminant, → Sorbent-Motivated Sorption.

Solvophoresis

A variant of diffusiophoresis. If a particle is immersed into a mixed solvent in which a concentration gradient exists, the particle tends to move in the direction of increasing concentration of the solvent component that is preferentially adsorbed onto its surface. *See* Reference [215].

Solvothermal Reaction

A chemical reaction in a solvent at, or near, supercritical conditions of pressure and temperature.

Somasundaran, Ponisseril (1939–)

An Indian surface and colloid chemist. Known for his work in flotation, adsorption, flocculation, electrokinetics, and mineral processing. Somasundaran and Fuerstenau are particularly known for their development of their multistage adsorption model that takes into account lateral interaction of surfactant molecules to form two-dimensional aggregates, termed hemi-micelles. He was the founding and long-time editor-in-chief of the journal Colloids and Surfaces. *See* Reference [216].

Sonication

A physical method of colloidal dispersion that uses ultrasound to create cavitation, which in turn provides the energy needed disperse particles, droplets, or bubbles. → Sonochemistry.

Sonochemistry

A chemical method of colloidal dispersion that uses ultrasound to create cavitation, which in turn creates highly localized regions in which chemical reactions take place under high temperature and pressure. → Sonication.

Soot

→ Lampblack.

Sorbate

A substance that becomes sorbed into an interface or another material or both. → Sorption.

Sorbent

The substrate into which or onto which a substance is sorbed or both. → Sorption.

Sorbent-Motivated Sorption

Sorption that occurs as a result of the affinity of the surface for a particular compound. Example: ion exchange. → Solvent-Motivated Sorption.

Sorption

A term used in a general sense to refer to either or both of the processes of adsorption and absorption.

Sorptive

Material that is present in one or both of the bulk phases bounding an interface and capable of becoming sorbed.

Sorptive Material	$\rightarrow$ Adsorptive Material.
Sparging	Sparging is the introduction of gas bubbles into a liquid through fine orifices. Example: sparging is used in mineral flotation cells.
Specific Increase in Viscosity	The relative viscosity minus unity. Also referred to as "specific viscosity" or "relative viscosity increment". *See* Table 4.
Specific Surface Area	$\rightarrow$ Surface Area.
Specific Viscosity	$\rightarrow$ Specific Increase in Viscosity.
Spherical Agglomeration	The process of separating particles from their suspension by selective wetting and agglomeration with a second, immiscible liquid. The second liquid preferentially wets the particles and causes particle adhesion (agglomeration flocculation) by capillary liquid bridges [217]. The process of agglomeration thus includes aggregation and coalescence. $\rightarrow$ Agglomeration.
Spinning Bubble Method	$\rightarrow$ Spinning Cylinder Method.
Spinning Cylinder Method	A method for determining the electrokinetic (zeta) potential of colloidal species, in which a bubble is held in a cylindrical cell that is spinning about its long axis. An electric field gradient is applied along the long axis of the cell to induce electrophoretic motion, and otherwise the method is essentially the same as particle micro-electrophoresis. In one such method, the inner surface of the cylinder is coated with polymer molecules in order to neutralize charges on that surface and therefore reduce or eliminate electro-osmotic flow inside the cylinder. Also termed Spinning Bubble Method, or the Spinning Electrophorometer Method. *See also* Table 16.
Spinning Drop Method	A method for determining surface or, more commonly, interfacial tension based on measuring the shape of a droplet (or bubble) suspended in the center of a horizontal, cylindrical tube filled with a liquid while the tube is spinning around its long axis. This method is particularly suited to the determination of very low interfacial tensions.
Spinning Electrophorometer Method	$\rightarrow$ Spinning Cylinder Method.
Spinning Oil	In the textile industry, a product applied to fibres in order to reduce friction, increase flexibility, and sometimes increase wetting, during combing and spinning operations. *See* Reference [41].
Spinning Solution Additive	In the textile industry, a surface-active product added to spinning solution to improve spinning operations [41]. Example: alkyl-sulfates. $\rightarrow$ Spinning Oil.

SPM	$\rightarrow$ Scanning Probe Microscopy.
Sponge Phase	A bicontinuous microemulsion phase. $\rightarrow$ Bicontinuous System, $\rightarrow$ Middle-Phase Microemulsion.
Spotting Agent	In the textile industry, a product applied to fibres or fabrics to remove stains [41]. Example: solvents formulated to contain surfactants such as alkylsulfates.
Spray	A dispersion of a liquid in a gas (aerosol of liquid droplets) in which the droplets have diameters greater than $10\,\mu$m. $\rightarrow$ Mist.
Spreadable Fats	A family of (usually) water-in-oil emulsion food products based on butter and margarine. They generally have fat contents of between 10 and 90% (w/w) and are solid at room temperature ($\sim$20 °C). They are categorized based on the amounts of components such as vegetable oils, animal fats and milk fat. Some examples of these "spreadable fats" include butter, butter mixture, margarine, low-fat spread, vegetable-fat spread, butterfat spread, low-calorie spread, and yellow-fat spread. Some very low fat content spreads, termed water-continuous spreads, have also been formulated as oil-in-water emulsions. *See* Reference [218].
Spreading	The tendency of a liquid to flow and form a film coating an interface, usually a solid or immiscible liquid surface, in an attempt to minimize interfacial free energy. Such a liquid forms a zero contact angle as measured through itself.
Spreading Coefficient	A measure of the tendency for a liquid to spread over a surface (usually of another liquid). It is -1 times the Gibbs free energy change for this process (the work of adhesion between the two phases minus the work of cohesion of the spreading liquid) so that spreading is thermodynamically favored if the spreading coefficient is greater than zero. In a gas-liquid system containing a potentially spreading liquid A, a substrate liquid L, and gas, the spreading coefficient is given by $S = \gamma^{\circ}_L - \gamma_{L/A} - \gamma^{\circ}_A$, where γ°_L and γ°_A are surface tensions and $\gamma_{L/A}$ is interfacial tension. For the spreading of liquid on a solid, a modified equation without the solid/gas tension is sometimes used. Other usages of the concept have involved terms such as the "spreading parameter" or "wetting power" [219], or spreading tension [41]. When equilibria at the interfaces are not achieved instantaneously, reference is frequently made to the initial spreading coefficient and final (equilibrium) spreading coefficient. If the initial spreading coefficient is positive and the final spreading coefficient negative, the system exhibits autophobicity. $\rightarrow$ Autophobing, $\rightarrow$ Entering Coefficient.
Spreading Parameter	$\rightarrow$ Spreading Coefficient.

Spreading Pressure	$\rightarrow$ Surface Pressure.
Spreading Tension	A synonym for spreading coefficient.
Spreading Wetting	The process of wetting in which a liquid, already in contact with a solid (or second, immiscible liquid) surface, spreads over the solid surface, thereby increasing the interfacial area of contact between them. The spreading is thermodynamically favored when the spreading coefficient is positive. $\rightarrow$ Adhesional Wetting, $\rightarrow$ Immersional Wetting, $\rightarrow$ Spreading Coefficient, $\rightarrow$ Wetting.
Spread Monolayer	$\rightarrow$ Spread Layer.
Spread Layer	The interfacial layer formed by an adsorbate when it becomes essentially completely adsorbed out of the bulk phase(s). If the layer is known to be one molecule thick, then the term "spread monolayer" is used.
Sputtering	The process by which atoms and/or ions are ejected from a target specimen due to particle bombardment.
Stable Foam	An oil- and gas-well drilling fluid foam that contains film-stabilizing additives, such as polymers or clays, and is pre-formed at the surface. $\rightarrow$ Air Drilling Fluid, $\rightarrow$ Foam Drilling Fluid, $\rightarrow$ Stiff Foam.
Stability	$\rightarrow$ Colloid Stability, $\rightarrow$ Thermodynamic Stability.
Stalagmometric Method	$\rightarrow$ Drop-Weight Method.
Standard Sedimentation Coefficient	A synonym for reduced sedimentation coefficient. $\rightarrow$ Sedimentation.
Static Coefficient of Friction	$\rightarrow$ Friction.
Static Foam Test	Any of several methods for assessing foam stability in which one measures the rate of collapse of a (static) column of foam [17, 71] Examples: Bartsch Test, Ross–Miles Test. $\rightarrow$ Dynamic Foam Test, $\rightarrow$ Foaminess.
Static Interfacial Tension	$\rightarrow$ Static Surface Tension.
Static Mixer	A device for mixing components in a solution or dispersion without moving mechanical elements. Stationary flow-guiding elements are built into a device, frequently a section of pipe, and they induce mixing and dispersion by repeatedly dividing and recombining partial streams of the flowing material. Also termed "motionless mixers".
Static Surface Tension	A synonym for "equilibrium surface tension" or "interfacial tension". $\rightarrow$ Surface Tension.

Staudinger, Hermann (1881–1965)	A German chemist known for his work in macromolecular, or polymer, chemistry by which he demonstrated that long-chain molecules could be created by covalently bonding together many small molecules, or monomers (he coined the terms "macromolecular" and "polymerization"). A prolific writer of books and other publications, Staudinger also discovered the ketenes. Eponyms include the "Staudinger–Mark–Houwink equation" (polymer solution viscosity). He won the Nobel prize in chemistry (1953) for his discoveries in the field of macromolecular chemistry. *See* Reference [220].
Staudinger–Mark–Houwink Equation	An empirical equation relating intrinsic viscosity to molecular mass for polymer solutions. *See* Table 11. This equation can used to determine viscosity-average molecular mass. Sometimes referred to as the "Mark–Houwink Equation".
STED	Stimulated Emission Depletion (Microscopy). $\rightarrow$ Super-Resolution Microscopy.
Stefan Flow	$\rightarrow$ Diffusiophoresis.
STEM	Scanning Transmission Electron Microscopy. $\rightarrow$ Electron Microscopy.
Steric Stabilization	The stabilization of dispersed species induced by the interaction (steric stabilization) of adsorbed polymer chains. The interaction creates a short-range, volume-restriction stabilization mechanism. Example: Adsorbed proteins stabilize the emulsified oil (fat) droplets in milk by steric stabilization. Also termed depletion stabilization. $\rightarrow$ Protection.
Stern Layer	The layer of ions in an electric double layer that, hydrated or not, lie adjacent to the surface (adsorbed ions). The rest of the electric double layer is often distinguished as the diffuse part, where assumptions, such as treating the ions as point charges, can more reasonably be made. $\rightarrow$ Electric Double Layer, $\rightarrow$ Helmholtz Double Layer.
Stern Layer Potential	$\rightarrow$ Inner Potential.
Stern, Otto (1888–1969)	A German physicist known for work on molecular beams, and his discovery of the magnetic moment of the proton, for which he received the Nobel prize (1943) in physics. Stern is best known to colloid scientists for contributions to understanding the structure of the electric double layer, particularly in extending the Helmholtz and Gouy-Chapman models to include counterions adsorbed on the surface. Such a layer of closely held counterions is known as a "Stern layer". *See* Reference [221].
Stern Potential	$\rightarrow$ Inner Potential.

STF	$\rightarrow$ Sculptured Thin Film.
Sticking Coefficient	The ratio of the rate of adsorption to the rate at which the adsorbate strikes the surface. This parameter indicates the likelihood that a species will adsorb.
Stiff Foam	An oil- and gas-well drilling fluid foam that contains film-stabilizing additives, such as polymers or clays; is pre-formed at the surface; and is more viscous than stable foam, having sufficient carrying capacity to remove drill cuttings from large diameter holes. Also termed "gel foam". $\rightarrow$ Air Drilling Fluid, $\rightarrow$ Foam Drilling Fluid, $\rightarrow$ Stable Foam.
Stimulated Emission Depletion (STED) Microscopy	$\rightarrow$ Super-Resolution Microscopy.
STM	$\rightarrow$ Scanning Tunneling Microscopy, $\rightarrow$ Scanning Probe Microscopy.
Stochastic Optical Reconstruction Microscopy (STORM)	$\rightarrow$ Super-Resolution Microscopy.
Stokes Diameter	The diameter of a sedimenting species determined from Stokes' law assuming a spherical shape. Also referred to as the "Stokes diameter" or (divided by a factor of 2) the "settling radius". $\rightarrow$ Equivalent Spherical Diameter.
Stokes, Sir George Gabriel (1819–1903)	An Irish-born U.K. mathematician and physicist known for his work on viscous fluids. Stokes' Law describes the velocity of small spheres sedimenting under the influence of an applied force (usually gravity). Stokes coined the term "fluorescence". Other eponyms include the "Navier-Stokes equation" (motion of viscous fluids), "Stokes-Einstein equation" (diffusion coefficient), and "Stokes", the unit of kinematic viscosity.
Stokes' Law	A relation giving the terminal settling velocity of a sphere as $2r^2\rho g/(9\eta)$, where r is the sphere radius, $\Delta\rho$ is the density difference between the phases, g is the gravitational constant, and η is the external-phase viscosity.
Stokes Settling	The settling of dispersed particles, droplets, or bubbles in accordance with Stokes Law. This usually applies for dilute dispersions of species having sizes smaller than about $100\,\mu m$ in diameter. In more concentrated dispersions, hydrodynamic interactions among the dispersed species cause a lower rate of sedimentation than predicted by Stokes Law. This phenomenon is termed hindered settling. Hindered settling can lead to a state in which all of the dispersed species sediment together at the same rate, regardless of their individual sizes. This phenomenon is termed zone

	settling. Under zone settling conditions a clear boundary will be apparent, between the sedimenting zone and the overlying supernatant solution.
Stopping Cross-Section	The energy loss of a particle impacting on a target per unit area density of target atoms expressed, for example, as $eV\,cm^2/atom$. Used in energetic ion analysis.
STORM	Stochastic Optical Reconstruction Microscopy. $\rightarrow$ Super-Resolution Microscopy.
Strain, Strain Rate	$\rightarrow$ Shear Rate.
Stratified Film	A fluid film in which several thicknesses can exist simultaneously and can persist for a significant amount of time. $\rightarrow$ Fluid Film.
Streaming Current	In electrokinetic motion, the current due to relative displacement of the part of the electric double layer beyond the shear plane; the electric current flowing in a streaming potential cell when the electrodes are short-circuited.
Streaming Potential	The potential difference at zero current created when liquid is made to flow through a porous medium. This method can be used to determine zeta potential. *See also* Table 16.
Streamline Flow	$\rightarrow$ Laminar Flow.
Steiner Angle	The $120°$ angle at which three foam films always come together to meet, at a Plateau border, in the dry limit. The equal angles results from equalization of the surface tension forces along the three liquid films.
Stress	$\rightarrow$ Shear Stress.
Stress Relaxation	$\rightarrow$ Viscoelastometer.
Strutt, John William (1842–1919)	A British physicist, better known to scientists as Lord Rayleigh, who made contributions to many sub-disciplines of physics (optics, acoustics, electromagnetism) and chemistry (thermodynamics, chemical physics). In colloid science he is best known for contributions to light scattering theory. The scattering of light by species whose size is much less than the wavelength of the incident light is referred to as "Rayleigh scattering". Other eponyms include the "Rayleigh ratio" (in light scattering), "Rayleigh Horn" (a fog horn), the "Rayleigh–Ritz approximation" and "Schroedinger–Rayleigh method" (quantum mechanics), and the "Rayleigh–Gans–Debye–Born approximation". He received the Nobel prize (1904) in physics for experiments on the density of gases. *See* References [222, 223].
Stirred-Media Mill	A device for reducing particle size by grinding. $\rightarrow$ Comminution.

Sublation   A flotation process in which the solute of interest becomes adsorbed on the surface of gas bubbles and is recovered in an upper layer of immiscible liquid.

Submicrometre Emulsion $\rightarrow$ Nanoemulsion, $\rightarrow$ Emulsion.

Submicron   An older particle size range distinction no longer in use. $\rightarrow$ Micrometre, $\rightarrow$ Micron.

Subphase $\rightarrow$ Substrate.

Subsidence   The process of sedimentation in which the settling of suspended particles results in a dense compaction, or coagulation, of particles in which liquid is squeezed out. Geologically, significant compaction of clay layers caused by lowering of the water table (dewatering).

Substrate   A material that underlies a surface or interface at which adsorption or other phenomena take place. It can be simply an underlying or supporting phase for a layer of distinct composition. A liquid phase that underlies an adsorption layer is called a "subphase".

Suction Pressure $\rightarrow$ Capillary Pressure.

Suds $\rightarrow$ Foam.

Suds Control Agents   Components in detergent formulations that act to stabilize or suppress sudsing (foaming). Examples: Alkylamine oxides can be added to promote or stabilize sudsing; silicones can be added to suppress sudsing (defoaming).

Sugden's Parachor $\rightarrow$ Parachor.

Sulfated Galactan   One of the kinds of polysaccharide structure that constitutes agar. Together with pyruvic acid acetal, the combination is sometimes referred to as "charged agar", or "agaropectin". $\rightarrow$ Agar.

Supercentrifuge $\rightarrow$ Centrifuge.

Superficial Density   An older term now replaced by the "Gibbs surface concentration", or simply, the "surface excess".

Superfluidifier $\rightarrow$ Plasticizer.

Superfluidizer $\rightarrow$ Plasticizer.

Superhardness $\rightarrow$ Hardness.

Super-Hydrophobic Surface   (SHS) Low energy surfaces, with a high degree of surface roughness, that can exhibit values of greater than $150°$ (usually with water). The increased contact angle, compared with that of a smooth surface, is thought to be due to the trapping of air in the spaces created by the rough surface. Also termed super-water-repellent surfaces. Synthetic examples of super-hydrophobicity include covers for solar cells, satellite dishes, and microfluidics, among others. → Lotus Effect.

Superplasticizer   → Plasticizer.

Super-Resolution Microscopy   Optical microscopy that can achieve resolution beyond the diffraction (Abbe) limit; that is, beyond the wavelength of visible light ($\sim$300–700 nm). One approach to super-resolution microscopy makes use of photo-activatable light-emitting fluorophore tags attached to large molecules like proteins, enabling them to be imaged with resolution of about 20 nm. Emerging such techniques include Stimulated Emission Depletion (STED) Microscopy, Photoactivated Localization Microscopy (PALM), Fluorescence Photoactivation Localization Microscopy (F-PALM), and Stochastic Optical Reconstruction Microscopy (STORM).

Superspreading   Refers to the wetting spreading of certain aqueous solutions on low energy, strongly hydrophobic surfaces, like polyethylene, on which water would not normally spread. Example: Superspreading occurs with solutions of some siloxane surfactants. Also termed superwetting.

Superwater   → Polywater.

Super Water-Reducer   → Plasticizer.

Superwetting   → Superspreading.

Surface   → Interface.

Surface-Active Agent   → Surfactant.

Surface Area   The area of a surface or interface, especially that between a dispersed and a continuous phase. The specific surface area is the total surface area divided by the mass of the appropriate phase.

Surface Charge   The fixed charge that is attached to, or part of, a colloidal species' surface and forms one layer in an electric double layer. There is thus a surface-charge density associated with the surface. → Electric Double Layer.

Surface-Charge Density   → Surface Charge.

Surface Colloids   → Solloids.

Surface Concentration   → Gibbs Surface.

Surface Conductivity   The excess conductivity, relative to the bulk solution, in a surface or interfacial layer per unit length. Also termed the "surface excess conductivity".

Surface Coverage   The ratio of the amount of adsorbed material to the monolayer capacity. The definition is the same for either of monolayer and multilayer adsorption.

Surface Dilational Modulus	$\rightarrow$ Film Elasticity.
Surface Dilational Viscosity	$\rightarrow$ Surface Viscosity.
Surface Elasticity	$\rightarrow$ Film Elasticity.
Surface Excess (Concentration)	$\rightarrow$ Gibbs Surface.
Surface Excess Conductivity	$\rightarrow$ Surface Conductivity.

Surface Excess Isotherm   A function relating, at constant temperature and pressure, the relative adsorption or reduced adsorption, or similar surface excess quantity, to the concentration of component in the equilibrium bulk phase.

Surface-Extended X-Ray Absorption Fine Structure Spectroscopy   (SEXAFS) $\rightarrow$ Extended X-Ray Absorption Fine Structure Spectroscopy.

Surface Fluidity   The inverse of the surface shear viscosity.

Surface Force Apparatus   (SFA) An apparatus used to obtain information about the nature and range of interparticle forces (the interparticle forces are not directly determined). In one widely used design, thin sheets of mica are cleaved and glued to cylindrical glass forms, which are then placed in cross-cylinder geometry with the mica sheets facing each other. The mica sheets can be prepared with coatings. An electrolyte or surfactant solution is placed between the cylinders, and the upper cylinder is moved toward the lower under a known load. An equilibrium separation distance (film thickness) of the order of nanometres, will be attained, which is measured by light interferometry. Measurements are made for varying loads, varying solution compositions, in different liquids, and with or without surface coatings. *See* References [224–226].

Surface Laser Light Scattering   (SLLS) An optical technique based on the surface capillary waves caused by thermal fluctuations of molecules at an interface. The capillary waves act as a diffraction grating and scatter incident laser light. At a given small angle the scattered light intensity contains information about the interfacial properties. The power spectrum of the scattered light intensity is measured, and the surface or interfacial tension and the bulk viscosity are inferred from capillary wave theory. *See* Reference [227].

Surface Layer	$\rightarrow$ Interfacial Layer.
Surface Layer of Adsorbent	$\rightarrow$ Adsorption Space.

Surface Load   In food colloids, refers to the amount of protein adsorbed at an interface. It is usually expressed in units of mg protein per m^2 of surface area. A monolayer of unfolded polypeptide chains produces a protein load of about $1\,\mathrm{mg\,m}^{-2}$, whereas protein loads involving aggregates or multilayers can reach tens of $\mathrm{mg\cdot m}^{-2}$ [228]. Also termed Protein Load.

Surface of Shear	$\rightarrow$ Shear Plane.
Surface of Tension	An imaginary boundary, having no thickness, at which surface or interfacial tension acts.
Surface Phenomena	Any phenomena whose effects are manifested at a surface separating two phases.
Surface Potential	The potential at the interface bounding two phases, that is, the difference in outer (Volta) potentials between the two phases. $\rightarrow$ Inner Potential, $\rightarrow$ Outer Potential, $\rightarrow$ Jump Potential.
Surface Potential Jump	$\rightarrow$ Jump Potential.
Surface Pressure	Actually an analog of pressure; the force per unit length exerted on a real or imaginary barrier separating an area of liquid or solid that is covered by a spreading substance from a clean area on the same liquid or solid. Also referred to as "spreading pressure".
Surface Rheology	$\rightarrow$ Surface Viscosity.
Surface Rheometer	$\rightarrow$ Surface Viscometer.
Surface Roughness	Any deviation of topography from an ideal, atomically smooth, planar surface. One measure of surface roughness is the rms deviation from the center-line average.
Surface Shear Viscosity	$\rightarrow$ Surface Viscosity.
Surface Tension	The contracting force per unit length around the perimeter of a surface is usually referred to as "surface tension" if the surface separates gas from liquid or solid phases, and as "interfacial tension" if the surface separates two nongaseous phases. Although not strictly defined the same way, surface tension can be expressed in units of energy per unit surface area. For practical purposes surface tension is frequently taken to reflect the change in surface free energy per unit increase in surface area. $\rightarrow$ Surface Work.
Surface Tension Methods	*See* Table 15.
Surface Viscometer	An instrument for determining surface rheological properties. One such type of instrument operates by rotating a ring or disk to apply shear in the plane of the interface, while maintaining the area constant. Alternatively, measurements are made by expanding or contracting the interface. $\rightarrow$ Canal Viscometer, $\rightarrow$ Torsional Viscometer.
Surface Viscosity	The two-dimensional analog of viscosity acting along the interface between two immiscible fluids. Also called "interfacial viscosity".

In fact, there are two kinds of surface viscosity: surface shear viscosity and surface dilational viscosity (or surface dilatational viscosity). Surface shear viscosity is the component that is analogous to three-dimensional shear viscosity: the rate of yielding of a layer of fluid caused by an applied stress. Surface dilational viscosity relates to the rate of area expansion and is expressed as the local gradient in surface tension per change in relative area per unit time. Any shear rate dependence (non-Newtonian behavior) falls under the subject of surface rheology. Although usually termed "surface viscosity" or rheology, especially when one fluid is a gas, the more general terminology is "surface" or "interfacial rheology". $\rightarrow$ Viscosity.

Surface Work

The work required to increase the area of the surface of tension. Under reversible, isothermal conditions the surface work (per unit surface area) equals the equilibrium, or static, surface tension.

Surfactant

Any substance that lowers the surface or interfacial tension of the medium in which it is dissolved. The substance does not have to be completely soluble and can lower surface or interfacial tension by spreading over the interface. Soaps (fatty acid salts containing at least eight carbon atoms) are surfactants. Detergents are surfactants, or surfactant mixtures, whose solutions have cleaning properties. Also referred to as "surface-active agents" or, for synthetic surfactants, "tensides". The term "surfactant" was originally a trademark of the General Aniline and Film Corp. and later released to the public domain [229]. The term "paraffin-chain salts" was used in the older literature [39]. In some usage, surfactants are defined as "molecules capable of associating to form micelles". *See also* Table 14.

Surfactant Affinity Difference

(SAD) Salager's surfactant affinity difference is used to model emulsion phase behaviour based on the chemical potentials of surfactant in aqueous and oil phases. The Hydrophilic-Lipophilic Deviation (HLD) is a dimensionless representation of SAD, given by $HLD = SAD/RT$. Either SAD or HLD values can be used to determine composition regions for which macroemulsions or microemulsions are likely to be stable, break or invert. Negative SAD or HLD values refer to Winsor Type 1 systems (oil-in-water), positive SAD or HLD values refer to Winsor Type II systems (water-in-oil), and $SAD = HLD = 0$ refers to Winsor Type III systems (most of the surfactant is in a middle phase with oil and water). *See* Reference [230].

Surfactant Effectiveness

The surface excess concentration of surfactant corresponding to saturation of the surface or interface. Example: One indicator of effectiveness is the maximum reduction in surface or interfacial tension achievable by a surfactant. This term has a different meaning from surfactant efficiency. *See* References [22, 23].

Surfactant Efficiency
The equilibrium solution surfactant concentration needed to achieve a specified level of adsorption at an interface. Example: One such measure of efficiency is the surfactant concentration needed to reduce the surface or interfacial tension by 20 mN/m from the value of the pure solvent(s). This term has a different meaning from surfactant effectiveness. *See* References [22, 23].

Surfactant Enhanced Aquifer Remediation
(SEAR) A remediation technology based on reservoir chemical flooding principles (micellar solubilization and/or low interfacial tension flooding) and applied to the treatment of NAPL-contaminated soils.

Surfactant Macromonomers
Hydrophobic monomers that also have surfactant character, also termed surfomers. Example: nonylphenoxypoly(etheroxy)ethyl acrylate. They are copolymerized with acrylamide to form hydrophobically associating polymers. → Hydrophobically Associating Polymers. *See* Reference [102].

Surfactant Tail
The lyophobic portion of a surfactant molecule. It is commonly a hydrocarbon chain containing eight or more carbon atoms. → Head Group.

Surfomers
→ Surfactant Macromonomers.

SUS
Saybolt Universal Seconds. → Saybolt Universal Viscosity.

Suspended Load
The particles that are picked-up and carried along by streams and rivers, maintained in suspension by turbulence. The colloidal sized fraction, particles of up to about 0.1 μm in diameter, that may be able to remain in suspension for considerable periods of time, even in quiescent waters is called the wash load.

Suspended Sediment
The insoluble particulate matter in natural water bodies such as rivers, lakes, and oceans. → Suspended Load.

Suspending Power
The ability of a detergent or detergent component to keep foreign material away from the solid material from which it has been removed to prevent redeposition. → Detergency, → Detergent.

Suspension
A system of solid particles dispersed in a liquid. Suspensions were previously referred to as "suspensoids", meaning suspension colloids. Aside from the obvious definition of a colloidal suspension, a number of operational definitions are common in industry, such as any dispersed matter that can be removed by a 0.45 μm nominal pore-size filter.

Suspension Concentrates
Highly concentrated (i.e., 30 to 60%) suspensions of fine particles, stabilized by surfactants, formulated to be easier than powders to

handle and disperse prior to use, produce no dust, and be easy to dilute. These surfactants are usually anionic, such as an alkyl diphenyl ether disulfonate. Also called Flowables. Example: pesticide suspension concentrates intended for spraying. → Emulsifiable Concentrates.

Suspension Effect

A finite potential, the Donnan potential, can exist between a suspension and its equilibrium solution. Also referred to as the "Pallmann" or "Wiegner effect".

Suspension Film

→ Liquid Film.

Suspo-Emulsion

A mixed colloidal dispersion in which a suspension is combined with an emulsion. Example: some kinds of emulsion-based paints.

SUV

→ Saybolt Universal Viscosity.

Svedberg

A unit of the sedimentation coefficient equal to 10^{-13} s.

Svedberg, Theodor (1884–1971)

A Swedish chemist known for his invention of the ultracentrifuge and its application to the study of proteins as well as his studies of Brownian motion. He discovered that proteins are macromolecules rather than associations of small molecules. Svedberg won the Nobel prize (1926) in chemistry for his work in colloid chemistry, particularly on Brownian motion. Eponyms include the "Svedberg" (a unit of the sedimentation coefficient) and the "Svedberg equation" (sedimentation). *See* References [231, 232].

Swamping Electrolyte

An excess of indifferent electrolyte that severely compresses electric double layers and minimizes the influence of electric charges borne by large molecules or dispersed colloidal species.

SWCNT

Single-Walled Carbon Nanotube. → Carbon Nanotube.

SWNT

Single-Walled Nanotube. → Carbon Nanotube.

Sweep Flocculation

A mechanism of aggregation or flocculation in which particles are enmeshed by a coagulant matrix. The particles are aggregated not due to charge neutralization but rather to enmeshment. Example: the rapid precipitation of a metal hydroxide from supersaturated solution where the settling fluffy hydroxide particles trap and enmesh other suspended particles. *See* Reference [136].

Swelling

Increase in volume associated with the uptake of liquid or gas by a solid or a gel.

Swelling Pressure

The pressure difference between a swelling material and the bulk of fluid being imbibed that is needed to prevent additional swelling. → Swelling.

Swirl Chamber Atomizer	Device for making aerosols of liquid droplets by pumping a liquid through a nozzle causing it to break-up into droplets at the orifice exit. These are used in many furnaces and combustion engines.
Swollen Micelle	$\rightarrow$ Micelle.
SXAPS	Soft X-Ray Appearance Potential Spectroscopy. $\rightarrow$ Appearance Potential Spectroscopy.
SXES	Soft X-Ray Emission Spectroscopy. $\rightarrow$ X-Ray Emission Spectroscopy.
Symmetric Film	$\rightarrow$ Film.
Synchrotron Radiation	Radiation caused by the acceleration of high energy electrons in a storage ring, or synchrotron.
Syndet	A synthetic detergent other than a soap.
Syndiotactic Polymer	$\rightarrow$ Atactic Polymer, $\rightarrow$ Tacticity.
Syneresis	The spontaneous shrinking of a colloidal dispersion due to the release and exudation of some liquid; frequently occurs in gels and foams but also occurs in flocculated suspensions. Mechanical syneresis refers to enhancing syneresis by the application of mechanical forces. Microsyneresis is a special case of syneresis in which the polymer molecules cluster together while retaining some of the original bulk gel structure. This process creates regions of free liquid within the gel network. *See* Reference [145].
Synergistic Electrolyte	$\rightarrow$ Critical Coagulation Concentration.
Syntactic Polymer	$\rightarrow$ Atactic Polymer, $\rightarrow$ Tacticity.
Szyszkowski Equation	An equation for estimating the surface tensions of aqueous solutions of various concentrations. *See* Table 8.

T

Tacticity

(1) Solutions and Dispersions. A tactic system has a regular or symmetric arrangement. Interactions or associations are sometimes referred to as occurring between identical (homotactic) or different (heterotactic) molecules or species in a given system.

(2) Surfaces. A tactic surface has a regular or symmetric surface structure. A homotactic surface has a homogeneous surface structure, as opposed to a heterotactic surface, which does not.

(3) Polymers. A tactic polymer has a regular or symmetric structure along its backbone. An atactic polymer has its substituents, or side chains, randomly distributed on either side of its backbone. An isotactic polymer has its substituents oriented on the same side of its backbone. A syndiotactic polymer has its substituents regularly alternating on each side of its backbone. → Atactic Polymer.

Tactoid

(1) In the destabilization of lyophilic colloids when coacervation occurs, the dispersed phase can initially separate into small, anisotropic droplets having shapes such as cylinders, called "tactoids". With concentrated colloids, droplets of dilute colloid can separate out within the concentrated colloid; these droplets are sometimes referred to as "negative tactoids".

(2) In clay suspensions the thin sheetlike or platelike particles can aggregate to form stacks of particles in face-to-face orientation, which are termed "tactoids".

Tall Oil

Fatty and resinous carboxylic acids obtained from the sulfate process used to obtain cellulose from softwood trees.

Tamman Temperature

The transition temperature at which atoms or molcules gain appreciable mobility and reactivity. For example, at the Tamman temperature the rate of sintering becomes significant. *See* Reference [9].

TAN

→ Total Acid Number.

Dictionary of Nanotechnology, Colloid and Interface Science. Laurier L. Schramm
Copyright © 2008 WILEY-VCH Verlag GmbH & Co. KGaA, Weinheim
ISBN: 978-3-527-32203-9

Tanaka, Toyoichi (1946–2000)	A Japanese physicist known for his development of smart gels, that can swell or contract by up to 1000 times in response to changes in physical conditions, such as temperature, illumination, solution properties, or an applied magnetic or electric field. $\rightarrow$ Smart Colloids. *See* Reference [233].
Taniguchi, Norio (1912–1999)	A Japanese engineering professor at Tokyo Science University who, in 1974, coined the term 'nano-technology' to describe processes on the order of a nanometre. Taniguchi is otherwise known for his work in high-precision machining and precision materials processing technologies. *See* References [234, 235]. $\rightarrow$ Drexler, (Kim) Eric, $\rightarrow$ Iijima, Sumio, and $\rightarrow$ Feynman, Richard P.
Tate's Law	An equation giving the mass of a droplet that forms and falls from a capillary tube of radius r, as the product of the surface tension times $2\pi r$. This equation is not very accurate, and significant corrections must be applied. $\rightarrow$ Drop-Weight Method.
Taylor Equation	An empirical equation for estimating the viscosity of an emulsion. *See* Table 11.
Taylor Turbulence Microscale	$\rightarrow$ Microturbulence.
TBN	$\rightarrow$ Total Base Number.
TDS	(1) Total Dissolved Solids. $\rightarrow$ Total Solids. (2) Thermal Desorption Spectroscopy. $\rightarrow$ Temperature-Programmed Reaction Spectroscopy.
Teller, Edward (1908–)	A Hungarian-born American physicist and physical chemist who is best known for his work on theoretical and nuclear physics, and as the "father of the hydrogen bomb". In earlier work he contributed to the areas of chemical bonding and adsorption. He is known in interface science for his contribution to the BET (Brunauer–Emmett–Teller) theory of multilayer adsorption. *See* References [236, 237].
TEM	Transmission Electron Microscopy. $\rightarrow$ Electron Microscopy.
Temkin Isotherm	An adsorption isotherm equation for heterogeneous surfaces based on a linear decrease in the enthalpy of adsorption with increasing surface coverage. Otherwise it is similar to the Langmuir adsorption isotherm. $\rightarrow$ Adsorption Isotherm, $\rightarrow$ Freundlich Isotherm.
Temperature-Programmed Reaction Spectroscopy	(TPRS) A surface technique in which thermally stimulated adsorbed species are desorbed from a surface to gain information about adsorbate-substrate bonding and about surface composition. This technique is destructive in that the heating drives off the

adsorbed species. Depending on whether the temperature rise is conducted quickly or slowly, two techniques are distinguished: flash desorption spectroscopy (FDS) and thermal desorption spectroscopy (TDS), also termed "temperature-programmed desorption spectroscopy" (TPDS). → Photon-Stimulated Desorption Spectroscopy.

Tenside	A synthetic surfactant. → Surfactant.
Tensiometer	A general term applied to any instrument that can be used to measure surface and interfacial tension.
Tertiary Electroviscous Effect	→ Electroviscous Effect.
Tertiary Oil Recovery	→ Enhanced Oil Recovery.
TFHBM	Thin-Film Helicoidal Bianisotropic Media. → Sculptured Thin Film.
Thermal Desorption Spectroscopy	(TDS) → Temperature-Programmed Reaction Spectroscopy.
Thermal Spraying	A method for creating nanostructured coatings in which a wire or powder is melted in a gas or plasma flame and then sprayed onto a surface.
Thermionic Work Function	The work needed to move an electron from the highest occupied level in a metal to a position outside the metal. Sometimes termed "Work Function".
Thermocapillary Diffusion	Temperature induced Marangoni flow. The movement of suspended drops or bubbles when subjected to a temperature gradient, caused by the resulting surface/interfacial tension gradient. *See*, for example, Reference [238].
Thermodynamic Stability	In colloid science, the terms "thermodynamically stable" and "metastable" mean that a system is in a state of equilibrium corresponding to a local minimum of free energy [4]. If several states of energy are accessible, the lowest is referred to as the "stable state" and the others are referred to as "metastable states"; unstable states are not at a local minimum. Most colloidal systems are metastable or unstable with respect to the separate bulk phases. → Colloid Stability, → Kinetic Stability.
Thermomechanical Writing	(Nanotechnology) → Scanning Probe Surface Patterning.
Thermophoresis	Particle motion, in an aerosol of solid particles, caused by non-uniform heating due to temperature gradients in the suspending gas. → Photophoresis.
Thermo-Shrinking Polymers	A kind of Smart Colloid involving dispersed colloidal microgel particles. Poly(N-isopropylacrylamide) microgels, for example,

exhibit a thermoreversible conformational transition in water at around 34 °C [13]. Below this temperature the microgel is highly swollen with water due to strong polymer-water interactions. Above this temperature polymer-polymer interactions become more important than the polymer-water interactions and the particles contract, causing the microgel to shrink and expel water. These changes are reversible, and can be quite rapid. → Smart Colloids.

Thermotropic Liquid Crystals	→ Mesomorphic Phase.
Thermotropic Mesomorphic Phase	→ Mesomorphic Phase.
Theta Temperature	The temperature at which a polymer solution exhibits ideal solution behavior. Also referred to as "Flory point" and "Flory temperature".
Thickness of the Electric Double Layer	→ Electric-Double-Layer Thickness.
Thin Film	→ Fluid Film.
Thin-Film Drainage	→ Film Drainage, → Fluid Film.
Thin-Film Helicoidal Bianisotropic Media	(TFHBM) → Sculptured Thin Film.
Thixotropic	Pseudoplastic flow that is time-dependent. At constant applied shear rate, viscosity gradually decreases, and in a flow curve hysteresis occurs. That is, after a given shear rate is applied and then reduced, it takes some time for the original dispersed species' alignments to be restored. The term thixotropy was coined by Freundlich and Bircumshaw in 1926 [239]. Example: quicksand. Thixotropy in gels is sometimes termed "reversible sol-gel transformation". If the original viscosity is not recovered after the shear is discontinued, the behaviour is termed "work softening".
Thomas Equation	An empirical equation for estimating the viscosity of a dispersion. *See* Table 11.
Thomson, William (1824–1907)	An Irish mathematician and physicist, also known as Lord Kelvin, who is famous for his work and his inventions in many areas, including electricity, thermodynamics, and marine engineering. Colloid and interface scientists use the Kelvin equation, which gives the vapour pressure above a drop of given radius and surface tension. In connection with his work on thermodynamics Kelvin proposed the absolute temperature scale which bears his name, having −273 degrees as absolute zero (the temperature at which the energy of motion of molecules reaches zero). Other eponyms include the "Joule-Thomson effect" (Kelvin coined the unit of

"Joule" for energy), "Kelvin electrometer", "Kelvin compass", and "Kelvin sounder". *See* Reference [240].

Three-Dimensional HLB System

(3D-HLB) An extension of the HLB system developed to handle silicone-containing surfactants and also emulsification in systems that may contain oil, water and silicone phases. The 3D-HLB system sets-up a right triangle with scales of 0 to 20, in which one coordinate is the conventional oil-in-water HLB value, one is the oil-in-silicone HLB (given by the % oil soluble divided by 5) and the other is the silicone-in-water HLB (given by the % water soluble divided by 5). Regions in the triangle can be "phase" mapped in terms of emulsion-forming tendency, such as oil-in-silicone versus water-in-silicone, oil-in-water versus silicone-in-water, silicone-in-oil versus water-in-oil, etc. The position of a particular surfactant in one of these regions predicts the type of emulsion that it will tend to form. *See* Reference [241]. → Hydrophile-Lipophile Balance.

Three-Phase Emulsion

→ Multiple Emulsion.

Three-Phase Separator

→ Separator.

Tight Emulsion

A petroleum industry term for a practically stable emulsion, in contrast to a less stable, or "loose", emulsion.

Tilt

Of a target: → Angle of Incidence.

Tilting-Plate Method

A means of determining the contact angle between a solid plate and the liquid into which it is immersed. The plate is adjusted to various angles until a flat (horizontal) meniscus is obtained, in which case the plate angle measured through the liquid yields the desired contact angle.

Tin Whisker

Micro- or nanoscale tendrils that can grow and extend outwards from a tin or tin-alloy surface. Freestanding tin whiskers are of interest in electronics and photonics because they can extend outwards from tin-plated surfaces, or from tin-silver-copper solder joints, and cause short-circuits. → Nanowhisker.

Tiselius Apparatus

An apparatus for the determination of electrophoretic mobilities. → Moving Boundary Electrophoresis.

Tiselius, Arne (Wilhelm Kaurin) (1902–1971)

A Swedish chemist known for his work in electrophoresis. He developed an improved electrophoresis method, using a cell called a Tiselius tube, for separating and studying protein mixtures. This improved method permitted the use of higher voltage gradients, and therefore better resolving power, than had previously been possible. He was awarded the Nobel prize (1948) in chemistry for his work in electrophoresis and adsorption analysis.

TOC	$\rightarrow$ Total Organic Carbon.
Toms Effect	The reduction in frictional drag that can result from the additions to a liquid of small concentrations of an additive, usually a polymer. The drag reduction may apply to, for example, spreading or pumping processes. Example: poly(ethylene)oxide is added to some fire-fighting foams in order to enhance their rate of spreading over a burning fuel. *See* References [242, 243].
Toothed Disc Mill	$\rightarrow$ Pin Mill.
Top-Down Processing	A subtractive, material removal process by which nanostructures are made from bulk materials. $\rightarrow$ Bottom-Up Processing.
Topochemical Reaction	A chemical reaction that can not be expressed stoichiometrically. Such reactions can occur only at certain locations on a molecule or only for certain molecular orientations.
Torsional Viscometer	An instrument for the qualitative determination of surface or interfacial viscosity in which a circular flat or double-cone bob, placed in the plane of and just contacting the surface or interface, is caused to oscillate. With the knife-edge supported by a torsion wire, the damping of oscillations is observed. Alternatively, in a two-dimensional analog of the concentric cylinder rheometer, constant rotational shear is applied and torque is measured.
Tortuosity	In porosimetry evaluations, experimental data tend to be interpreted in terms of a model in which the porous medium is taken to comprise a bundle of cylindrical pores having radius r. If the Young-Laplace equation is then applied to the data, an effective value of r can be calculated, even though this model ignores the real distribution of irregular channels. The calculated r value is sometimes considered to represent the radius of an equivalent cylinder or, alternatively, is termed the "tortuosity".
Total Acid Number	(TAN) The acid number expresses the amount of base (potassium hydroxide) that will react with a given amount of material in a standardized titration procedure. A large acid number indicates a high concentration of acids in the original material, usually including natural surfactant precursors. A commonly measured property of crude oils.
Total Base Number	(TBN) The base number expresses the amount of acid that will react with a given amount of material in a standardized titration procedure. A large base number indicates a high concentration of bases in the original material.
Total Dissolved Solids	(TDS) $\rightarrow$ Total Solids.

Total Organic Carbon	(TOC) A measure of the amount of Natural Organic Matter (NOM) in (usually) a natural water sample. The soluble fraction is termed Dissolved Organic Carbon (DOC). Example: a natural water sample may contain both insoluble (cell debris) and soluble (humic substances) materials.
Total Porosity	→ Porosity.
Total Potential Energy of Interaction	→ Gibbs Energy of Interaction.
Total Solids	(TS) The concentration of dissolved and suspended impurities in water. The total solids is equal to the sum of the total dissolved solids (TDS) and the total suspended solids (TSS). For most natural, fresh waters, the TDS is mostly made up of dissolved salts and is typically in the range of 50–1000 mg/l. In sea water, the TDS level is about 35 g/l [244]. The total concentration of insoluble species in water is the total suspended solids (TSS). For many natural waters the TSS is typically in the range of 10–20 mg/l but can be much higher [244].
Total Suspended Solids	(TSS) → Total Solids.
TPRS	→ Temperature-Programmed Reaction Spectroscopy.
Tracer Diffusion Coefficient	The diffusion coefficient of an isotopically labeled species. Usually taken to be equal to the diffusion coefficient of the corresponding unlabeled species. → Diffusion Coefficient.
Transitional Pore	An older term now replaced by "mesopore". → Pore.
Transition Velocity	The flow velocity in a pipe or stirred vessel that corresponds to a transition from laminar to turbulent flow conditions, or vice versa. → Critical Deposition Velocity.
Transitive Nanoparticle	A nanoparticle exhibiting a size-related intensive property that differs significantly from that observed in fine particles or bulk materials [32]. This term is used when the material has properties that emerge only on the nanoscale, that is, whose behavior does not smoothly or simply extrapolate from the bulk. → Non-Transitive Nanoparticle.
Transmission Electron Microscopy	(TEM) → Electron Microscopy.
Traube's Rule	A generalization for homologous series of organic compounds of type $R(CH_2)_nX$, that, for each incremental CH_2 group, the concentration of molecules required to produce a specified surface tension decreases by a factor of about 3. In adsorption Traube's rule is that a polar adsorbent will preferentially adsorb the most polar component from a nonpolar solution, and conversely, a nonpolar

adsorbent will preferentially adsorb the least polar component from a polar solution.

Treater	A vessel used for the breaking of emulsions and the consequent removal of solids and water (BS&W). Emulsion breaking can be accomplished through some combination of thermal, electrical, chemical, or mechanical methods. A treater might be applied to break an emulsion and separate solids and water that could not be removed in a separator.
Tribolever	A synonym for the Nano-Triboscope. → Nanotribology.
Tribology	The science of friction and lubrication and wear. Tribological wear occurs as a result of interacting surfaces in relative motion. → Nanotribology.
Triple Emulsion	→ Multiple Emulsion.
Triple-Phase Emulsion	→ Multiple Emulsion.
True Colloid	→ Pseudocolloid.
TS	→ Total Solids
TSS	Total Suspended Solids. → Total Solids.
TU	Turbidity Unit. → Nephelometric Turbidity Unit.
Turbidimetry	→ Turbidity.
Turbidity	The property of dispersions that causes a reduction in the transparency of the continuous phase due to light scattering and absorption. Turbidity is a function of the size and concentration of the dispersed species. The turbidity coefficient is simply the extinction coefficient in the Beer-Lambert equation for absorbance when light scattering rather than absorbance proper is being studied (hence turbidimetry). → Nephelometry.
Turbidity Unit	(TU) A unit of measurement in light scattering, equal to the turbidity of a standard suspension containing $1\,\mu g/mL$ of silica particles. This has largely been replaced by the nephelometric turbidity unit, and associated standard suspension. → Nephelometric Turbidity Unit.
Turbulent Flow	A condition of flow in which all components of a fluid passing a certain point do not follow the same path. Turbulent flow refers to flow that is not laminar, or streamline.
Turner Velocity	The minimum, or critical, velocity of flowing gas in a vertically oriented pipe needed to suspend a droplet of liquid. Developed for petroleum industry gas wells flowing under pressure the Turner

equation relates the minimum gas flow velocity to an empirical constant, surface tension, and the densities of the liquid and gas phases. *See also* References [143, 245], $\rightarrow$ Soap Sticks.

Tyndall Beam $\rightarrow$ Tyndall Scattering.

Tyndall Effect   When a sufficiently dilute dispersion of small particles or droplets is viewed directly against an illuminating light source, it appears completely transparent. In contrast, when the same dispersion is viewed from the side (at a right angle to the illuminating beam), and against a dark background, the dispersion appears somewhat turbid and blue-white in colour. The scattered light is due to Tyndall scattering, and the optical effect is referred to as the "Tyndall effect". $\rightarrow$ Tyndall Scattering.

Tyndall, John (1820–1893)   An Irish-born British physicist known to colloid science for contributions in the field of aerosols, including the scattering of light by colloidal particles (the Tyndall effect), and the explanation of the blue colour of the sky in terms of such light scattering (as opposed to, for example, colour contributed by oxygen in the atmosphere). Other eponyms include "Tyndall beam" (the light scattered by colloidal particles) and "Tyndallization" (a process like pasteurization). Tyndall is also known for contributions in other areas of physics (magnetism, acoustics) and chemistry (thermodynamics, infrared spectroscopy). *See* References [246, 247].

Tyndall Scattering   A process that produces a colored beam of light scattered by uniform dispersion of particles whose size approaches the wavelength of the incident light. The scattered light is referred to as a "Tyndall beam". The wavelength of scattered light varies with the angle of observation (Tyndall spectra), and this feature allows particle size to be calculated. $\rightarrow$ Higher-Order Tyndall Spectra.

U

Ubbelohde Viscometer	→ Capillary Viscometer.
Ultracentrifuge	→ Centrifuge.
Ultrafiltrate	→ Filtration, → Ultrafiltration.
Ultrafiltration	A separation process somewhat like dialysis in which a colloidal dispersion is separated from a noncolloidal solution by a semipermeable membrane; that is, a membrane permeable to all species except the colloidal-sized ones. Here an applied pressure (rather than osmotic pressure) across the membrane drives the separation. As in dialysis, the solution containing the colloidal species is referred to as the "retentate" or "dialysis residue". However, the solution that is free of colloidal species is referred to as "ultrafiltrate" rather than "dialysate", because the composition is usually different from that produced by dialysis. Also referred to as "hyperfiltration" or "reverse osmosis" (although the latter term refers to "sub-colloidal solutes"). → Dialysis, → Filtration, → Reverse Osmosis.
Ultrafine Aerosol	→ Nanoaerosol.
Ultrafine Bubble	→ Nanobubble.
Ultrafine Droplet	→ Nanodroplet.
Ultrafine Emulsion	→ Nanoemulsion, → Emulsion.
Ultrafine Grinding	→ Attrition.
Ultrafine Particle	→ Nanoparticle.
Ultramicroscope	An optical microscope that uses dark-field illumination to make visible extremely small (submicrometre-sized) particles or droplets. Also termed "Dark-Field Microscope". → Dark-Field Illumination.

Dictionary of Nanotechnology, Colloid and Interface Science. Laurier L. Schramm
Copyright © 2008 WILEY-VCH Verlag GmbH & Co. KGaA, Weinheim
ISBN: 978-3-527-32203-9

Ultrasonic Dispersion	The use of ultrasound waves to achieve or aid in the dispersion of particles or droplets.
Ultrasound Vibration Potential	(UVP) An electroacoustical method for determining the electrokinetic potential of colloidal species which are detected by the electric field generated when the species are made to move by an imposed ultrasonic field. This method can be applied to many dispersions, including W/O emulsions that do not transmit light. For low potentials, the UVP can be quantitatively related to the electrophoretic mobility. Also called "Colloid Vibration Potential (CVP)" and, where the current is measured, "Colloid Vibration Current (CVC)." $\rightarrow$ Electrokinetic Sonic Amplitude, *see* Table 16, and *see* Reference [45, 96].
Ultraviolet Photoelectron Spectroscopy	(UPS) $\rightarrow$ Photoelectron Spectroscopy.
Ultraviolet Photoemission Spectroscopy	(UPS) $\rightarrow$ Photoelectron Spectroscopy.
Unactivated Adsorption	Physisorption; that is, adsorption for which there is no activation energy barrier to be overcome, as opposed to activated adsorption, or chemisorption, for which an activation energy barrier must be overcome. $\rightarrow$ Chemisorption, $\rightarrow$ Physisorption.
Upper Consolute Temperature	$\rightarrow$ Consolute Temperature.
Upper Critical Solution Temperature	$\rightarrow$ Consolute Temperature.
Upper-Phase Microemulsion	A microemulsion with a high oil content that is stable while in contact with a bulk water phase, and in laboratory tube or bottle tests it tends to be situated at the top of the tube above the water phase. For chlorinated organic liquids, which are more dense than water, the oil will be the top phase rather than the bottom. $\rightarrow$ Microemulsion, $\rightarrow$ Winsor-Type Emulsions.
Upper Plastic Limit	$\rightarrow$ Liquid Limit.
UPS	Ultraviolet Photoelectron Spectroscopy or Ultraviolet Photoemission Spectroscopy. $\rightarrow$ Photoelectron Spectroscopy.
U.S. Bureau of Mines Wettability Test	$\rightarrow$ Wettability Index.
UVP	$\rightarrow$ Ultrasound Vibration Potential.

V

Vacuole	Membrane-bound sacs (essentially large vesicles) that are contained in cells. Vacuoles are involved in intracellular nutrient storage and digestion, and in the release of cellular waste products. They exist in both plants and animals. $\rightarrow$ Vesicle.
van der Waals Adsorption	An older term now replaced by "physical adsorption" or "physisorption". $\rightarrow$ Chemisorption.
van der Waals Forces	The attractive interaction forces between any two bodies of finite mass. van der Waals forces include the Keesom orientation forces between permanent dipoles, Debye induction forces between dipoles and induced dipoles, and London (dispersion) forces between two induced dipoles. Also referred to as Lifshitz–van der Waals forces. $\rightarrow$ Dispersion Forces.
van der Waals–Hamaker Constant	$\rightarrow$ Hamaker Constant.
van der Waals, Johannes Diderik (1837–1923)	A Dutch physicist who worked on molecular physics and thermodynamics. He is known for the van der Waals equation of state for gases (pressure-volume-temperature) and for the general term "van der Waals forces" which refers collectively to the (other than ionic) attractive forces between molecules. van der Waals received the Nobel prize in physics (1910) for his work on the equation of state. *See* Reference [248].
van der Waals Bearing	A mechanical component in nanotechnology comprising a cylinder-and-sleeve bearing in which the mating surfaces are covered with close-packed atoms selected to provide repulsive van der Waals forces and therefore provide frictionless rotation. Example: a drive-shaft, in which the axle is less than 2 nm in diameter, made from diamond and where the mating cylinder and sleeve surfaces are both covered with fluorine atoms. *See* Reference [249].

Dictionary of Nanotechnology, Colloid and Interface Science. Laurier L. Schramm
Copyright © 2008 WILEY-VCH Verlag GmbH & Co. KGaA, Weinheim
ISBN: 978-3-527-32203-9

van't Hoff, Jacobus
(Henricus) (1852–1911)

A Dutch chemist who (with Ostwald and Arrhenius) helped establish the discipline of physical chemistry. He is known for his work in the concepts of energy, thermodynamics, and dilute solutions and osmosis. He is known in organic chemistry for originating (with Le Bel) the concept of tetrahedrally coordinated carbon, and for laying the foundation of modern stereochemistry. Eponyms include the "van't Hoff equation" (osmotic pressure) and "Vanthoffite" (a natural sodium-magnesium sulfate). van't Hoff received the first Nobel prize in Chemistry (1901) for his discovery of the laws of chemical dynamics and osmotic pressure in solutions. *See* Reference [250].

van't Hoff's Law

The relation specifying that the osmotic pressure of a solution equals the gas pressure the solute would exert if it were an ideal gas occupying the same volume as the solution.

Varnish

A protective surface coating product comprising an oil, resin, a solvent and/or a thinner. Once applied the solvent evaporates and the surface film hardens as a result of chemical reactions (curing) among the remaining components. Varnishes do not contain pigment so the final surface film is transparent with little or no colour. → Lacquer.

Vegetable-Fat Spread

→ Spreadable Fats.

Vehicle

In surface coatings or paint, → Binder.

Velocity Gradient

A parameter that indicates the intensity of mixing. It is a function of the power input, the reactor volume, and the fluid viscosity. Higher velocity gradients are used in coagulation where the goal is to disperse the coagulant to the particle surfaces. Lower velocity gradients are used in flocculation where the goal is particle collisions and aggregation, and higher gradients would break up flocs.

Venturi Atomizer

→ Air-Blast Atomizer.

Versator

A device used for deaerating liquid systems, such as emulsions, and operates on the principle of centrifugally generating a thin film of the liquid with high shear and exposing the thin film to vacuum.

Verwey, Evert (Johannes
Willem) (1905–1981)

A Dutch chemist known for physical chemical research in the industrial sphere. Together with J.Th.G. Overbeek, he developed a theory of the stability of colloidal particles that could explain the Schulze–Hardy rule. Independently developed also by Derjaguin and Landau the theory became famous as the "DLVO theory", an acronym constructed from the last names of these four scientists.

Very Coarse Sand

→ Sand, *see* Table 12.

Very Fine Sand

→ Sand, *see* Table 12.

Vesicle	A droplet characterized by the presence at its surface of a lipid bimolecular film (bilayer) or series of concentric bilayers. A vesicle can be single or multi-lamellar and stabilized by natural or synthetic surfactants. Vesicles made from phospholipid bilayers are called liposomes. Nanoscale liposomes are termed nanosomes. Vesicles made from synthetic nonionic surfactants are called niosomes. Both liposomes and niosomes are used as carriers in drugs and in personal care products. *See* Reference [9]. → Bimolecular Film, → Colloidosome, → Vacuole.
VI	→ Viscosity Index.
Vicinal Water	Water whose structure is modified somewhat (increased hydrogen bonding, increased specific heat and viscosity) due to the presence of a neighbouring solid surface. The interfacially modified water can apparently extend over distances of at least 20 molecular layers. *See* References [251, 252].
Virial Equation	An equation that includes a power series of terms of increasing order in the independent variable. Each term has associated with it a coefficient termed its "virial coefficient". An example of a virial equation is the nonideal gas law when written as a power series. The second virial coefficient describes the first order of deviations from nonideality.
Viscoelastic	A liquid (or solid) with both viscous and elastic properties. A viscoelastic liquid will deform and flow under the influence of an applied shear stress, but when the stress is removed the liquid will slowly recover from some of the deformation. → Boger Fluid.
Viscoelastometer	An instrument for studying viscoelastic fluids. Viscoelastometers can be used to apply a constant shear stress so that the resulting deformation can be determined (creep curve), or to apply a sudden deformation and determine the stress needed to maintain the deformation (stress relaxation).
Viscoelectric Constant	A reflection of the increase in viscosity of a liquid due to the presence of an electric field. It is given by the increase in viscosity divided by the viscosity in the absence of an electric field, and divided also by the square of the electric-field gradient.
Viscometer	Any instrument employed in the measurement of viscosity. In most cases the term is applied to instruments capable of measuring only Newtonian viscosity and not capable of non-Newtonian measurements. → Rheometer.
Viscosimeter	→ Viscometer.
Viscosity	A measure of the resistance of a liquid to flow. It is properly the coefficient of viscosity, and expresses the proportionality between

shear stress and shear rate in Newton's law of viscosity. For variations, *see* Table 4. Many equations have been used to predict the viscosities of colloidal dispersions; *see* Table 11.

Viscosity-Average Molecular Mass	Molecular mass determined on the basis of viscosity measurements coupled with an empirical equation such as the Staudinger-Mark-Houwink equation.
Viscosity Index	(VI) An indicator used to describe the temperature dependence of a fluid's kinematic viscosity. In terms of relative changes, a higher viscosity index represents a smaller decrease in viscosity with increasing temperature. The calculation of viscosity index involves the use of published look-up tables such as those in ASTM Designation D 2270-93. The viscosity index is primarily a petroleum industry measure.
Viscosity Index Improver	Agents (usually high-molar mass polymers) that increase the viscosity of a liquid over its useful temperature range. Such polymers are necessary in the formulation of multi-grade engine oils.
Viscosity Modifier	An additive, such as a polymer, that reduces a fluid's viscosity variations with temperature.
Viscosity Number	→ Reduced Viscosity.
Viscosity Ratio	→ Relative Viscosity.
Void	The cavities and/or passages inside a solid. Interconnected void spaces are usually termed pores or pore spaces. → Pore, → Porosity, → Porous Medium, → Void Ratio.
Void Ratio	The ratio of the volume of void space to the volume of solid in a material, such as a soil, sediment, or rock. → Porosity.
Vold, Marjorie (Jean Young) (1913–1991)	A pioneering American woman colloid chemist who is known for her work in the phase behaviour and kinetics of surfactant solutions, liquid crystals, and suspensions. Although she also worked in industry she was among the first wave of women professors in academia in North America. She is particularly known as the first author of the textbook *Colloid Chemistry: The Science of Large Molecules, Small Particles and Surfaces*, which was one of the most important books on colloid science when it was published in 1964. *See* References [253, 254].
Volta, Alessandro (Giuseppe Antonio Anastasio), Count (1745–1827)	An Italian physicist who invented the electrophorus (a device to produce static electric charge), demonstrated that electricity could originate in the junction of different metals, developed the electromotive series, and developed the forerunner of the battery: the Voltaic pile. The fundamental unit of electric potential, the volt, is

named for Volta. Volta also discovered methane by isolating and examining the properties of marsh gas.

Volta Potential $\rightarrow$ Outer Potential.

Votator   A continuous-process device for rapidly changing the temperature of a liquid system. Liquid enters the device, is spread in a thin film over a heat-exchanging surface, is then removed from the surface by wall scrapers, and then exits the device through an outlet.

Wagner Equation	An equation for predicting the conductivity of dispersions. *See* Table 10.
Wasan, Darsh T. (1938–)	Indian-born American chemical and colloid and interface engineer. Known for work in particle science and technology, emulsions, foams and thin films. Long-time editor of the Journal of Colloid and Interface Science.
Washburn Equation	An equation describing the extent of displacement of one fluid by another in a capillary tube or cylindrical pore in a porous medium. If h is the depth of penetration of invading fluid and dh/dt is the rate of penetration, then $dh/dt = \gamma r \cos \theta/(4\eta h)$, where γ is the interfacial tension, r is the capillary radius, θ is the contact angle, and η is the viscosity of the invading fluid. It is used in the evaluation of porosimetry data and can be used to provide information about contact angles, capillary radii, and pore radii, depending on the experiments conducted.
Washing De-Inking	$\rightarrow$ De-Inking.
Wash Load	$\rightarrow$ Suspended Load.
Water-Continuous Spread	$\rightarrow$ Spreadable Fats.
Water II	$\rightarrow$ Polywater.
Water Lock Effect	$\rightarrow$ Jamin Effect.
Water-Reducer	$\rightarrow$ Plasticizer.
Wavelength Dispersive X-Ray Spectroscopy	(WDXS) $\rightarrow$ Energy Dispersive X-Ray Spectroscopy.
WDXS	Wavelength Dispersive X-ray Spectroscopy. $\rightarrow$ Energy Dispersive X-Ray Spectroscopy.

Dictionary of Nanotechnology, Colloid and Interface Science. Laurier L. Schramm
Copyright © 2008 WILEY-VCH Verlag GmbH & Co. KGaA, Weinheim
ISBN: 978-3-527-32203-9

Wet	→ Wettability.
Wet Foam	→ Gas Emulsion.
Wet Nanotechnology	The nanoscale technology of chemical and/or biological systems in an aqueous (water-based) environment. In contrast, dry nanotechnology refers to "dry" or vacuum situations, and is sometimes used to refer to "dry" molecular manufacturing, nanomachines, and other nanodevices.
Wet Oil	An oil containing free water or emulsified water.
Wettability	A qualitative term referring to the water- or oil-preferring nature of surfaces, such as mineral surfaces. Example: The flow of emulsions in porous media is influenced by the wetting state of the walls of pores and throats through which the emulsion must travel. Wettability can be determined by direct measurement of contact angles, or inferred from measurements of fluid imbibition or relative permeabilities. Several conventions for describing wettability values exist. → Amott Test, → Contact Angle, → Wettability Index, → Wetting.
Wettability Index	A measure of wettability based on the U.S. Bureau of Mines (USBM) wettability test in which the wettability index (W) is determined as the logarithm of the ratio of areas under the capillary pressure curves for both increasing and decreasing saturation of the wetting phase. Complete oil-wetting occurs for $W = -\infty$ (in practice about -1.5), and complete water-wetting occurs for $W = \infty$ (in practice about 1.0). Another wettability index is derived from the Amott-Harvey test. *See also* Reference [36], → Wettability, → Amott Test.
Wetting	A general term referring to one or more of the following specific kinds of wetting: adhesional wetting, spreading wetting, and immersional wetting. Frequently used to denote that the contact angle between a liquid and a solid is essentially zero and there is spontaneous spreading of the liquid over the solid. Nonwetting, on the other hand, is frequently used to denote the case where the contact angle is greater than 90° so that the liquid rolls up into droplets. → Draves Wetting Test, → Contact Angle, → Wettability.
Wetting Coefficient	In the equilibrium of a system of solid, gas, and liquid, the wetting coefficient, k, is given as $k = (\gamma_{sg} - \gamma_{sl})/\gamma_{lg}$, where γ is the interfacial tension and the subscripts g, l, and s refer to gas, liquid, and solid phases, respectively, at the interfaces. Complete wetting occurs for $k \geq 1$ and nonwetting for $k \leq 1$.
Wetting Film	→ Liquid Film.

Wetting Hysteresis	A phenomenon in which the work involved in introducing a solid surface into a liquid is different from the work involved in withdrawing it. → Contact-Angle Hysteresis.
Wetting Power	→ Spreading Coefficient.
Wetting Tension	The work done on a system during the process of immersional wetting, expressed per unit area of the phase being immersionally wetted. → Immersional Wetting.
WfK Fabric	→ Krefeld Fabric.
Whey	A term in dairy processing referring to the dilute oil-in-water emulsion that separates from the coagulated portion, or curd, in cheese-making. → Curds.
Whey-Off	A term in dairy processing referring to any unwanted process in which whey separates from a product.
Whipping Agent	→ Foaming Agent.
Wicking	The flow of liquid into a porous medium due to capillary forces.
Wiegner Effect	→ Suspension Effect.
Wien Effect	The increase in electrical conductivity of an electrolyte solution at very high applied electric field strengths. The effect occurs when the ions in solution move fast enough to cause reductions in the electrophoretic relaxation and retardation effects. See Reference [255].
Wiener Equation	For predicting the relative permittivity of dispersions. See Table 9.
Wilhelmy, Ludwig (Ferdinand) (1812–1864)	A German chemist and physicist who conducted pioneering studies in chemical kinetics. He is remembered to colloid and interface scientists for his introduction of the concept that the surface tension of a liquid can be determined from the maximum force required to pull a small plate vertically from the surface of a liquid, now known as the "Wilhelmy Plate Method". See Reference [256].
Wilhelmy Plate Method	A method for determining surface or interfacial tension based on measuring the force needed to pull an inert plate, held vertically, through an interface. Also termed the Wilhelmy slide method. → du Noüy Ring Method.
Williams Landel Ferry Equation	(WLF Equation) An equation describing the effect of temperature on polymer solution viscosity. A common approximate form gives the relative viscosity of a polymer solution (η_{Rel}) as: $\ln \eta_{Rel} = -17.4(T - T_g)/\{51.6 + (T - T_g)\}$, where T_g is the glass-transition temperature for the polymer. See also Table 11.

Wilson, Charles (Thomson Rees) (1869–1959)	A Scottish meterologist known for his studies of clouds and their properties. He received the 1927 Nobel prize in physics for his invention of the Wilson cloud chamber.
Wilson Cloud Chamber	A device used to prepare aerosols of liquid droplets, or clouds, by the use of sudden expansion to induce supersaturation in a gas. This in turn causes aerosol droplet formation by nucleation on particles in the chamber. Charged particles, such as alpha or beta particles leave a "trail" of mist droplets as they move through the chamber. *See* Reference [257].
Winding Oil	In the textile industry, a product applied to yarns to reduce friction and increase flexibility to improve operations such as knitting [41]. Example: oils formulated with surfactants to make them emulsifiable in water.
Wine Tears	The layer of droplets that can appear at the top of the meniscus in a container of an alcohol-water solution (e.g., wine). The pumping action that draws liquid up through the meniscus is a result of the evaporation of alcohol from a thin region at the top of the layer, raising the surface tension and causing liquid to rise from the bulk into the layer (Marangoni flow). In the layer, droplets form, and they are drawn down by the force of gravity. → Marangoni Effect.
Winslow Effect	→ Electro-Rheological Colloids.
Winsor-Type Emulsions	Several categories of microemulsions that refer to equilibrium phase behaviors and distinguish, for example, the number of phases that can be in equilibrium and the nature of the continuous phase [258]. They are denoted as Winsor Type I (oil-in-water), Type II (water-in-oil), Type III (most of the surfactant is in a middle phase with oil and water), and Type IV (water, oil, and surfactant are all present in a single phase). The Winsor Type III system is sometimes referred to as a "middle-phase microemulsion", and the Type IV system is often referred to simply as a "microemulsion". An advantage of the Winsor category system is that it is independent of the density of the oil phase and can lead to less ambiguity than do the lower-phase or upper-phase microemulsion type terminology. Nelson-type emulsions are similarly identified, but with different type numbers.
WLF Equation	→ Williams Landel Ferry Equation.
W/O	Abbreviation for a water-dispersed-in-oil emulsion.
Work Function	→ Thermionic Work Function.
Work Hardening	→ Rheopexy.
Work Softening	→ Thixotropic.

Work of Adhesion	The energy of attraction between molecules in a phase. Defined as the work per unit area done on a system when two phases meeting at an interface of unit area are separated reversibly to form unit areas of new interfaces of each with a third phase.
Work of Cohesion	The work per unit area done on a system when a body of a phase is separated reversibly to form two bodies of the phase, each forming unit areas of new interfaces with a third phase.
Work of Immersional Wetting	$\rightarrow$ Wetting Tension.
Work of Separation	Synonym for the work of adhesion.
Work of Spreading	This term is equivalent to the spreading coefficient.
W/O/A	Abbreviation for a thin fluid film of oil between water and air phases. $\rightarrow$ Fluid Film.
W/O/W	(1) In multiple emulsions: Abbreviation for a water-dispersed-in-oil-dispersed-in-water multiple emulsion. Here the oil droplets have water droplets dispersed within them, and the oil droplets themselves are dispersed in water forming the continuous phase. (2) In fluid films: Abbreviation for a thin fluid film of oil in a water phase. Note the possibility of confusion with the multiple emulsion convention. $\rightarrow$ Fluid Film.

X

Xerogel	A xerogel is not a gel but rather is used with reference to a dried-out, possibly open, structure that was a gel. Also spelled "zerogel". Example: silica gel. $\rightarrow$ Aerogel.
XES	$\rightarrow$ X-Ray Emission Spectroscopy.
XPS	X-Ray Photoelectron Spectroscopy. $\rightarrow$ Photoelectron Spectroscopy.
X-Ray Diffraction	(XRD) A technique in which the scattering of X-rays by a crystal lattice is measured and used to determine the crystal's structure.
X-Ray Emission Spectroscopy	(XES) A technique used for the determination of surface composition by scanning the surface with an X-ray or electron beam. The beam ionizes surface atoms by ejecting inner-shell electrons. Electron transfer from outer electron shells will result in the emission of energy as characteristic X-rays. The spectrum of the emitted X-rays is then determined. A derived technique is soft X-ray emission spectroscopy (SXES). $\rightarrow$ Auger Electron Spectroscopy, *see* Table 7.
X-Ray Photoelectron Spectroscopy	(XPS) $\rightarrow$ Photoelectron Spectroscopy.
XRD	$\rightarrow$ X-Ray Diffraction.

Dictionary of Nanotechnology, Colloid and Interface Science. Laurier L. Schramm
Copyright © 2008 WILEY-VCH Verlag GmbH & Co. KGaA, Weinheim
ISBN: 978-3-527-32203-9

Y

Yellow-Fat Spread	→ Spreadable Fats.
Yield Stress	For some fluids, the shear rate (flow) remains at zero until a threshold shear stress is reached; this threshold is termed the "yield stress". Beyond the yield stress, flow begins. Also termed the "yield value". Some descriptions appropriate to various yield stresses are given in Table 20.
Yield Value	→ Yield Stress.
Young–Laplace Equation	The fundamental relationship giving the pressure difference across a curved interface in terms of the surface or interfacial tension and the principal radii of curvature. In the special case of a spherical interface, the pressure difference is equal to twice the surface (or interfacial) tension divided by the radius of curvature. Also referred to as the "equation of capillarity".
Young Modulus	→ Hooke's Law.
Young's Equation	A fundamental relationship giving the balance of forces at a point of three-phase contact. For a gas-liquid-solid system Young's equation is $\gamma_{SL} + \gamma_{LG} \cos \theta = \gamma_{SG}$, where γ_{SL}, γ_{LG}, and γ_{SG} are interfacial tensions between solid-liquid, liquid-gas, and solid-gas, respectively, and θ is the contact angle of the liquid with the solid, measured through the liquid.
Young, Thomas (1773–1829)	A British physicist and physician, Young is probably best known for his work on the wave nature of light. He is also known to colloid and interface scientists for his work on surface tension (Young–Laplace equation), contact angles (Young's equation), and elastic materials (Young's modulus).

Dictionary of Nanotechnology, Colloid and Interface Science. Laurier L. Schramm
Copyright © 2008 WILEY-VCH Verlag GmbH & Co. KGaA, Weinheim
ISBN: 978-3-527-32203-9

Z

Zahn Viscosity

A parameter intended to approximate fluid viscosity and measured by a specific kind of orifice viscometer. The measurement unit is Zahn seconds.

Z-average Mean

The hydrodynamic (equivalent spherical) diameter of a particle in a liquid, determined by dynamic light scattering. → Equivalent Spherical Diameter.

Zeolites

A class of aluminosilicate minerals having large cavities in their crystal structures. These cavities allow ion exchange of large ions and also can permit the size-selective passage of organic molecules. Zeolites are used as ion exchangers, molecular sieves, and catalysts.

Zerogel

→ Xerogel.

Zero Point of Charge

→ Point of Zero Charge, → Electrocapillarity.

Zeta Potential

Strictly called the "electrokinetic potential", the zeta potential refers to the potential drop across the mobile part of the electric double layer. Any species undergoing electrokinetic motion, such as electrophoresis, moves with a certain immobile part of the electric double layer, which is assumed to be distinguished from the mobile part by a distinct plane, the shear plane. The zeta potential is the potential at that plane and is calculated from measured electrokinetic mobilities (e.g., electrophoretic mobility) or potentials (e.g., sedimentation potential) by using one of a number of available theories.

Zimm, Bruno (Hasbrouck) (1920–2005)

An American physical chemist known for contributions in the areas of light scattering, and polymer and protein solutions. He worked particu-larly on light scattering theory where particle dimensions are not small compared with the wavelength of the light, in which Zimm Plots are used. (→ Mie, Gustav).

Dictionary of Nanotechnology, Colloid and Interface Science. Laurier L. Schramm
Copyright © 2008 WILEY-VCH Verlag GmbH & Co. KGaA, Weinheim
ISBN: 978-3-527-32203-9

Zimm Plot	A way of plotting scattering data (static light scattering, small-angle neutron scattering or small-angle X-ray scattering) for species such as polymers. By extrapolating the plotted data to zero concentration and analyzing the slope and intercept, the mass average molar mass, second virial coefficient, and the radius of gyration can be determined. This can be compared with a Guinier plot which is frequently used for solid particles (colloids).
Zisman Plot	$\rightarrow$ Critical Surface Tension of Wetting.
Zone Electrophoresis	A method for the separation of charged colloidal particles or large molecules. An electric-field gradient is imposed along a supporting medium, such as a gel, and a sample of mixture to be separated is applied to one end of the supporting medium. As electrophoretic motion occurs, regions of different components separate out along the direction of the electric-field gradient according to the different electrophoretic mobilities of the components. $\rightarrow$ Isoelectric Focussing.
Zone Settling	$\rightarrow$ Stokes Settling.
Zsigmondy, Richard Adolf (1865–1929)	An Austrian chemist who became interested in colloidal gold sols and coloured glasses. He developed the ultramicroscope, which allows one to "observe" colloidal particles in a dispersion by their scattering of light (Tyndall scattering). He received the Nobel prize (1925) in chemistry for his work in colloid chemistry.
Zwitterionic Surfactant	A surfactant molecule that contains negatively and positively charged groups. Example: lauramidopropylbetaine, $C_{11}H_{23}CONH(CH_2)_3N^+(CH_3)_2CH_2COO^-$, at neutral and alkaline solution pH. $\rightarrow$ Amphoteric Surfactant.

Appendix 1: List of Tables

Table 1 Types of colloidal dispersion.

Dispersed phase	Dispersion medium	Name	Examples
Liquid	Gas	Aerosol of liquid droplets	Fog, mist
Solid	Gas	Aerosol of solid particles	Smoke, dust
Gas	Liquid	Foam	Soap suds
Liquid	Liquid	Emulsion	Milk, mayonnaise
Solid	Liquid	Sol, suspension	Inks, gels, bacteria in water
Gas	Solid	Solid foam	Polystyrene foam
Liquid	Solid	Solid emulsion	Opal
Solid	Solid	Solid suspension	Ruby-stained glass, pearl

Table 2 Illustrative listing of nanoterms[1]

Artificial Atoms	Fullerene	Molecular Manufacturing
BioNEMS	Incidental Nanoparticle	Molecular Fabrication
Bottom-Up Processing	Inorganic Nanotube	Molecular Engineering
Buckminsterfullerene	Inorganic Fullerene-Like Material	Molecular Wire
Buckyball, Buckytube	Lab-on-a-Chip	Multi-Walled Nanotubes
Chemical Nanosensor	Nanoactive	Nano Approach
Chemosynthesis	Nanoaerosol	Nanoarchitecture
Coaxial Nanowire	Nano air	Nanoarray
Columnar Thin Film	Mechanosynthesis	Nanoassembler
Composite Clay Nanostructures	Mesoscopic Atoms	Nanobalance
Dip Pen Nanolithography	Microfluidics	Nanobeam
Electron Pump	Molecular Circuitry	Nanobelt
Electrospinning	Molecular Nanotechnology	Nanobiosensor
Engineered Nanoparticles	Molecular Machine	Nanobot

(*Continued*)

Dictionary of Nanotechnology, Colloid and Interface Science. Laurier L. Schramm
Copyright © 2008 WILEY-VCH Verlag GmbH & Co. KGaA, Weinheim
ISBN: 978-3-527-32203-9

Table 2 (*Continued*)

Nanobottle	Nanogeoscience	Nanoscopic
Nanobubble	Nanografting	Nanosensor
Nanocage	Nanohardness	Nanoshell
Nanocapsule	Nanohole	Nanosize
Nanocell	Nanohorn	Nanosizing
Nanoceramic	Nanoimprinting	Nanosheets
Nanochip	Nanoimprint Lithography	Nanosome
Nanochondrion	Nanoindenter	Nanosphere
Nanocluster	Nanolithography	Nanostructure
Nanocoating	Nanology	Nanosuspension
Nanocomposite	Nanomachine	Nanotechnology
Nanocomputer	Nanomanipulator	Nano Test Tube
Nanocore	Nanomanufacturing	Nanotribology
Nanocrane	Nanomaterial	Nanotube
Nanocrystal	Nanomechanical	Nanotweezers
Nanocube	Nanomesh	Nanovalve
Nanodevice	Nanometre	Nanowhisker
Nano-Differential Mobility	Nanomotor	Nanowire
Analysis	Nanonail	Nested Nanoparticle
Nanorhelogy	Nano-Onion	Non-Transitive Nanoparticle
Nanodispersion	Nanoparticle	Nucleation Aerosol
Nanodisassembler	Nanopen	Photonic Band Gap Structure
Nanodot	Nanopharmaceutical	Positional Synthesis
Nanodroplet	Nanophase	Quantum Bit, Qubit, Qbit
Nanoelectrochemical Patterning	Nanophotolithography	Quantum Computer
Nanoelectromechanical System	Nanophotonics	Quantum Dot
Nanoelectronic	Nanopipette	Quantum Well
Nanoemulsion	Nanoplotter	Quantum Wire
Nanoencapsulation	Nanoporous	Scanning Probe Surface Patterning
Nanoengineering	Nanopowder	Sculptured Nematic Thin Film
Nanofabrication	Nanoprobe	Sculptured Thin Film
Nanofatigue	Nanoreplicator	Single-Electron Device
Nanofibre	Nanoresist	Single-Walled Nanotubes
Nanofibril	Nanorhelogy	Thermomechanical Writing
Nanofilter	Nanoribbon	Thin-Film Helicoidal Bianisotropic
Nanoflask	Nanorobot	Media
Nanofluidics	Nanorod	Transitive Nanoparticle
Nanofoam	Nanorope	Ultrafine Droplet
Nanogap	Nanoscale	Ultrafine Grinding
Nanogel	Nanoscience	Ultrafine Particle

1) Not all terms are specifically defined in this book. Many "nano" terms simply refer to the base term in the context of the nanometre scale. For example, nanolithography is lithography at the nanometre scale.

Table 3 Some decimal-multiple prefixes for units in science and technology.

SI Prefix	SI Prefix	Meaning	Synonym
yotta	Y	10^{24}	septillion
zetta	Z	10^{21}	sextillion
exa	E	10^{18}	quintrillion
peta	P	10^{15}	quadrillion, quad, quartrillion
tera	T	10^{12}	trillion
giga	G	10^{9}	billion, thousand million, milliard
mega	M	10^{6}	million
kilo	k	10^{3}	thousand
hecto	h	10^{2}	
deka	da	10^{1}	
		10^{0}	
deci	d	10^{-1}	
centi	c	10^{-2}	hundredth
milli	m	10^{-3}	thousandth
micro	μ	10^{-6}	millionth
nano	n	10^{-9}	billionth
pico	p	10^{-12}	trillionth
femto	f	10^{-15}	quadrillionth
atto	a	10^{-18}	quintrillionth
zepto	z	10^{-21}	sextillionth
yocto	y	10^{-24}	septillionth

Table 4 Glossary of viscosities.

Term	Symbol	Explanation
Absolute viscosity	η	$\eta = \tau/\dot{\gamma}$ and can be traced to fundamental units independent of the type of instrument
Apparent viscosity	η_{APP}	$\eta_{APP} = \tau/\dot{\gamma}$ but as determined for a non-Newtonian fluid, usually by a method suitable only for Newtonian fluids.
Differential viscosity	η_D	$\eta_D = d\,\tau/d\,\dot{\gamma}$
Inherent viscosity	η_{Inh}	$\eta_{Inh} = C^{-1}\ln(\eta/\eta_o)$
Intrinsic viscosity	$[\eta]$	$[\eta] = \lim_{c \to 0} \lim_{\dot{\gamma} \to 0} \eta_{SP}/C$
		$[\eta] = \lim_{c \to 0} \lim_{\dot{\gamma} \to 0} (1/C)\ln \eta_{Rel}$
Kinematic viscosity	η_K	$\eta_K = \eta/\rho$
Reduced viscosity	η_{Red}	$\eta_{Red} = \eta_{SP}/C$
Relative viscosity	η_{Rel}	$\eta_{Rel} = \eta/\eta_o$
Specific increase in viscosity	η_{SP}	$\eta_{SP} = \eta_{Rel} - 1$

Symbols: ρ is the density of the bulk fluid or dispersion, η_o is the viscosity of the pure solvent or dispersion medium, and C is the dispersed-phase concentration (usually volume fraction).

Table 5 Some classifications for atmospheric aerosols of liquid droplets.

Classification	Lower-upper size limit (µm)
Fog	0.5–30
Cloud	2–100
Drizzle	100
Rain	100–about 5000

Source: *See* References [259, 260].

Table 6 Some approximate sizes for atmospheric aerosols of solid particles.

Classification	Approximate lower-upper sizes (µm)
By size-range:	
Aitken Nuclei	<0.2
Large Nuclei	0.2–2.0
Giant Nuclei	>2.0
By type:	
Fume, Smoke	0.01–1
Dust	0.5–100
Ash	1–500

Source: References [261].

Table 7 Some surface techniques and their acronyms.

Acronym	Technique	Information
AES	Auger electron spectroscopy	Surface composition
AEAPS	Auger electron appearance potential spectroscopy	Surface composition
AFM	Atomic force microscopy	Surface morphology
APD	Azimuthal photoelectron diffraction	Surface structure
APS	Appearance potential spectroscopy	Surface composition
APXPS	Appearance potential X-ray photoemission spectroscopy	Surface composition
ARPES	Angle-resolved photoemission spectroscopy	Surface structure
CELS	Characteristic energy-loss spectroscopy	Adsorbed composition
DFM	Dynamic force microscopy	Surface morphology
EDXS	Energy dispersive X-ray spectroscopy	Surface chemical analysis
EELS	Electron energy-loss spectroscopy	Adsorbed composition
EID	Electron-impact desorption spectroscopy	Surface, adsorbed composition
EIS	Electron-impact spectroscopy	Adsorbed composition
ELEED	Elastic low-energy electron diffraction	Surface structure
ELS	Energy-loss spectroscopy	Adsorbed composition
EPMA	Electron probe microanalysis	Surface composition
EPR	Electron paramagnetic resonance spectroscopy	Surface structure

Table 7 (*Continued*)

Acronym	Technique	Information
ESCA	Electron spectroscopy for chemical analysis	Surface composition
ESD	Electron-stimulated desorption	Surface structure, composition
EXAFS	Extended X-ray absorption fine structure spectroscopy	Surface atom packing
FDS	Flash desorption spectroscopy	Surface, adsorbed composition
FEM	Field emission microscopy	Surface structure
FFM	Friction force microscopy	Microscale friction and lubrication
FIM	Field ion microscopy	Surface structure
HEED	High-energy electron diffraction	Surface structure
HEIS	High-energy ion scattering	Surface composition, structure
HREELS	High-resolution electron energy-loss spectroscopy	Surface structure, composition
ILS	Ionization-loss spectroscopy	Surface composition
INS	Ion-neutralization spectroscopy	Surface, adsorbed electron structure
IRAS	Infrared reflection-absorption spectroscopy	Surface structure, composition
ISS	Ion-scattering spectroscopy	Surface composition
LEED	Low-energy electron diffraction	Surface structure
LEIS	Low-energy ion scattering	Surface composition, structure
MBS	Molecular beam spectroscopy	Surface reaction kinetics
MEIS	Medium-energy ion scattering	Surface composition, structure
NPD	Normal photoelectron diffraction	Surface structure
NSOM	Near-field scanning optical microscopy	Surface morphology
OSEE	Optically stimulated exoelectron emission spectroscopy	Adsorbed composition
PAS	Photoacoustic spectroscopy	Surface, adsorbed vibrational states
PES	Photoelectron spectroscopy	Surface composition
PhD	Photoelectron diffraction	Surface structure
PSD	Photon-stimulated desorption	Surface structure, composition
RHEED	Reflection high-energy electron diffraction	Surface structure, composition
SANS	Small angle neutron scattering	Surface features, adsorption layers
SEM	Scanning electron microscopy	Surface morphology, composition
SEXAFS	Surface-extended X-ray absorption fine structure	Surface structure, composition
SFA	Surface force apparatus	Short-range force between particles
SHEED	Scanning high-energy electron diffraction	Surface heterogeneity
SHG	Optical second harmonic generation	Surface dynamics and reactions
SIMS	Secondary ion mass spectroscopy	Surface composition
SLLS	Surface laser light scattering	Surface or interfacial tension
STM	Scanning tunnelling microscopy	Surface morphology
SXAPS	Soft X-ray appearance potential spectroscopy	Surface composition
SXES	Soft X-ray emission spectroscopy	Surface composition
TDS	Thermal desorption spectroscopy	Surface, adsorbed composition
TEM	Transmission electron microscopy	Surface structure
TPDS	Temperature-programmed desorption spectroscopy	Surface, adsorbed composition
TPRS	Temperature-programmed reaction spectroscopy	Surface, adsorbed composition
UPS	Ultraviolet photoelectron (photoemission) spectroscopy	Surface structure, composition
WDXS	Wavelength dispersive X-ray spectroscopy	Surface chemical analysis
XES	X-ray emission spectroscopy	Surface composition
XPS	X-ray photoelectron spectroscopy	Surface structure, composition
XRD	X-ray diffraction	Surface structure

Source: References [5, 155, 158, 262, 263].

Table 8 Equations for predicting surface and interfacial tensions.[1]

Name	Equation	Explanation
Antonow's rule	$\gamma_{12} = \gamma_1 - \gamma_2$	For liquid/liquid or liquid/solid interfacial tensions
Eötvös equation	$\gamma^\circ = k(T_c - T)(M/\rho)^{-2/3}$	
Fowkes equation	$\gamma_{12} = \gamma_1 + \gamma_2 - 2\sqrt{(\gamma_1^{\,d}\gamma_2^{\,d})}$	γ^d represents dispersion components of surface tensions
Girifalco-Good equation	$\gamma_{12} = \gamma_1 + \gamma_2 - 2a\,(\sqrt{\gamma_{12}})$	
Ideal mixing rule	$\gamma^\circ_{\,s} = \gamma^\circ_{\,1}X_1 + \gamma^\circ_{\,2}X_2$	For solutions, s, of two components, 1 and 2, having mole fractions X_1 and X_2
Sugden's parachor equation	$\gamma^\circ = (P\Delta\rho/M)^4$	P is an empirical constant, the parachor
Szyszkowski equation	$\gamma^\circ = \gamma^\circ_{\,0}\{1 - b\ln((1 + C)/a)\}$	Aqueous solutions of concentration C

1) Most of these equations are semi-empirical and some are specific to particular types of colloidal dispersion. Therefore, some of the equations appear somewhat contradictory.
Symbols: a and b are empirical constants, M is molecular mass, T_c is critical temperature, ρ is density, and $\gamma^\circ_{\,0}$ is the surface tension of pure solvent.

Table 9 Equations for predicting relative permittivities of dispersions.[1]

Name	Equation	Explanation
Böttcher equation	$(\in - \in_C)/(3\in) = \phi[(\in_D - \in_C)/(\in_D + 2\in)]$	
Bruggeman equation	$(\in - \in_D)/(\in_C - \in_D) = (1 - \phi)(\in/\in_C)^{1/3}$	For high ϕ
Fradkina equation	$\in = \in_C(1 + 3\phi)$	For W/O emulsions
Hanai equation	$(\in/\in_C) = 1/(1 - \phi)^3$	For low-frequency measurement
Wiener equation	$(\in - \in_C)/(\in + 2\in_C) = \phi(\in_D - \in_C)/(\in_D + 2\in_C)$	For $\phi \ll 1$

1) Most of these equations are semi-empirical and some are specific to particular types of colloidal dispersion. Therefore, some of the equations appear somewhat contradictory.
Symbols: $\in$ is the relative permittivity of the dispersion, $\in_D$ is the relative permittivity of the dispersed phase, $\in_C$ is the relative permittivity of the continuous phase, and a is a constant.
Source: Reference [20].

Table 10 Equations for predicting conductivities of dispersions.[1]

Name	Equation	Explanation
Bruggeman equation	$(\kappa - \kappa_D)(\kappa_C/\kappa)^{1/3} = (1 - \phi)(\kappa_C - \kappa_D)$ $(\kappa/\kappa_C) = (1 - \phi)^{3/2}$	When $\kappa_C \gg \kappa_D$
Fricke equation	$\kappa = \kappa_C - [\phi(\kappa - \kappa_D)/3(1 - \phi)]\sum\kappa_C/[\kappa_C(1 - a_i) + \kappa_D a_i]$	a_i is a series of further terms
Hanai equation	$(\kappa/\kappa_C) = 3\in(\in - \in_C)/[(\in_D + 2\in)(\in_D - \in_C)]$ $(\kappa/\kappa_C) = 1/(1 - \phi)^3$	For high-frequency measurement For low-frequency measurement
Lemlich equation	$(\kappa/\kappa_C) = (1 - \phi)/3$	For foams
Wagner equation	$(\kappa - \kappa_C)/(\kappa + 2\kappa_C) = \phi(\kappa_D - \kappa_C)/(\kappa_D + 2\kappa_C)$	

1) Most of these equations are semi-empirical and some are specific to particular types of colloidal dispersion. Therefore, some of the equations appear somewhat contradictory.
Symbols: κ is the conductivity of the dispersion, κ_D is the conductivity of the dispersed phase, κ_C is the conductivity of the continuous phase, ϕ is the volume fraction of the dispersed phase, and a is a constant.
Source: Reference [20].

Table 11 Equations for predicting viscosities of dispersions.[1]

Name	Equation	Explanation
Brinkman equation	$\eta = \eta_o(1-\phi)^{-2.5}$	For suspensions
Broughton-Squires equation	$\eta = \eta_o \exp(a_1\phi + a_2)$	For emulsions
Carreau equation	$\eta = \eta_N (1 + (\tau_r\dot{\gamma}^2)^{(n-1)/2}$	For polymer solutions, $\eta_N =$ low shear Newtonian viscosity, $\tau_r =$ rotational relaxation time, $\dot{\gamma} =$ shear rate $n =$ power law exponent
Eilers equation	$\eta = \eta_o (1 + 2.5\phi + 4.94\phi^2 + 8.78\phi^3)$	Emulsions of viscous oils
Einstein equation	$\eta = \eta_o (1 + 2.5\phi)$	For $\phi < 0.02$
Guth-Gold-Simha equation	$\eta = \eta_o (1 + 2.5\phi + 14.1\phi^2)$	For $\phi < 0.06$
Hatschek equation	$\eta = \eta_o (1/\{1-\phi^{1/a}\})$	For emulsions, $\phi > 0.5$
Krieger-Dougherty equation	$\eta = \eta_0 (1-\phi/\phi_{max})^{2.5\phi max}$	For concentrated suspensions, $\phi_{max} =$ maximum packing fraction
Mooney equation	$\eta = \eta_o \exp(2.5\phi/\{1-a\phi\})$	For emulsions
Modified Mooney equation	$\eta = \eta_o \exp(K_1\phi^{DE}/\{1-K_2\phi^{DE}\})$	$\phi^{DE} =$ volume fraction of dispersed oil and dispersed water
Oliver-Ward equation	$\eta = \eta_o(1 + a\phi + a^2\phi^2 + a^3\phi^3 + \ldots)$	For spheres
Richardson equation	$\eta = \eta_o \exp(a\phi)$	For emulsions
Sibree equation	$\eta = \eta_o(1/\{1-(a\phi)^{1/a}\})$	For emulsions
Simha equation	$\eta = \eta_o(1 + 32\phi/\{15p\pi\})$	Anisometric particle suspensions
Staudinger-Mark-Houwink equation	$[\eta] = KM^a$	For polymer solutions
Taylor equation	$\eta = \eta_o(1 + 2.5\phi\{(\eta_D + 0.4\eta_o)/(\eta_D + 0.4\eta_o)\})$	Emulsions of viscous oils
Thomas equation	$\eta = \eta_o(1 + 2.5\phi + 10.05\phi^2 + 0.00273 \exp(16.6\phi))$	For suspensions
Williams Landel FerryEquation	$\ln \eta_{Rel} = -17.4(T-T_g)/\{51.6 + (T-T_g)\}$	For polymer solutions

1) Most of these equations are semi-empirical and some are specific to particular types of colloidal dispersion. Therefore, some of the equations appear somewhat contradictory.

Symbols: η_o is the viscosity of the pure solvent or continuous phase, η_D is the viscosity of the dispersed phase, K and a are empirical constants, p is the ratio of particle thickness to particle diameter (and must be small), and T_g is the glass-transition temperature (polymer).

Table 12 Some particle size classifications extending upwards from the classical colloidal domain.

Classification	Wentworth	Soil Sci. Soc. Amer.
Clays	0–3.9	0–2.0
Silt	3.9–62.5	2.0–50
Very fine sand	62.5–125	50–100
Fine sand	125–250	100–250
Medium sand	250–500	250–500
Coarse sand	500–1000	500–1000
Very coarse sand	1000–2000	1000–2000
Gravel-granule	2000–4000	2000–80,000

Note: All values are lower and upper size limits in micrometres.
Source: References [264, 265].

Table 13 Equations for predicting critical micelle concentrations.

Name	Equation	Explanation	Reference
La Mesa equation	$|C_{red} - 1| = |1 - T_{red}|^{1.74}$	Ionic surfactants to 60 °C	[266]
La Mesa equation	$|C_{red} - 1| = |1 - P_{red}|^{2.8}$	Ionic surfactants to 200 MPa	[187]
Stasiuk–Schramm equation	$|C_{red} - 1| = 853|1 - T_{red}|^{3.54}$	Ionic surfactants to 180 °C	[267]
Stasiuk–Schramm equation	$|C_{red} - 1| = 2.25 \times 10^4|1 - T_{red}|^{5.80}$	Amphoteric surfactants to 180 °C	[188]

Symbols: Reduced variables are $C_{red} = C/C_o$, $P_{red} = P/P_o$, and $T_{red} = T/T_o$ where C is critical micelle concentration in mol/L, P is pressure, and T is temperature in K; Subscript o indicates value of the property at the minimum critical micelle concentration.

Table 14 Some carboxylic (fatty) acid names.[1]

Carbon	Name	Formula	Synonyms
C_8	Octanoic acid	$CH_3(CH_2)_6CO_2H$	Caprylic acid
C_9	Nonanoic acid	$CH_3(CH_2)_7CO_2H$	n-Nonylic acid, Pelargonic acid
C_{10}	Decanoic acid	$CH_3(CH_2)_8CO_2H$	Capric acid, n-Decylic acid
C_{11}	Hendecanoic acid	$CH_3(CH_2)_9CO_2H$	Undecanoic acid, Undecylic acid
C_{12}	Dodecanoic acid	$CH_3(CH_2)_{10}CO_2H$	Lauric acid, Undecane-1-carboxylic acid
C_{13}	Tridecanoic acid	$CH_3(CH_2)_{11}CO_2H$	Tridecylic acid
C_{14}	Tetradecanoic acid	$CH_3(CH_2)_{12}CO_2H$	Myristic acid
C_{14}	9-Tetradecenoic acid (*cis*)	$CH_3(CH_2)_3CH{:}CH(CH_2)_7CO_2H$	Myristoleic acid
C_{15}	Pentadecanoic acid	$CH_3(CH_2)_{13}CO_2H$	Pentadecylic acid
C_{16}	Hexadecanoic acid	$CH_3(CH_2)_{14}CO_2H$	Palmitic acid
C_{16}	9-Hexadecenoic acid (*cis*)	$CH_3(CH_2)_5CH{:}CH(CH_2)_7CO_2H$	Palmitoleic acid
C_{17}	Heptadecanoic acid	$CH_3(CH_2)_{15}CO_2H$	Margaric acid
C_{18}	Octadecanoic acid	$CH_3(CH_2)_{16}CO_2H$	Stearic acid
C_{18}	9-Octadecenoic acid (*cis*)	$CH_3(CH_2)_7CH{:}CH(CH_2)_7CO_2H$	Oleic acid
C_{19}	Nonadecanoic acid	$CH_3(CH_2)_{17}CO_2H$	n-Nonadecylic acid
C_{20}	Eicosanoic acid	$CH_3(CH_2)_{18}CO_2H$	Arachidic acid, Eicosoic acid
C_{21}	Heneicosanoic acid	$CH_3(CH_2)_{19}CO_2H$	

1) The simple salts of these acids, or soaps, are significantly surface active in water from about C_{10} through C_{20}. Their names generally follow the following example: the sodium salt of dodecanoic acid is sodium dodecanoate; in older terminology, the sodium salt of lauric acid is sodium laurate.

Table 15 Methods for determining surface and interfacial tensions.

Name	Acronym	Synonym	Surface tension	Interfacial tension
Capillary rise			√	
Captive bubble method			√	
Captive drop method				√
Critical surface tension of wetting			√	
Drop volume			√	√
Drop weight			√	
du Noüy ring			√	√
Imbedded disc retraction	IDR			√
Imbedded fibre retraction	IFR			√
Maximum bubble pressure			√	
Maximum droplet pressure				√
Oscillating jet		Elliptical jet		√
Pendant bubble			√	
Pendant drop			√	√
Sessile bubble			√	
Sessile drop			√	√
Solidification front		Freezing front	√	√
Spinning drop				√
Surface laser light scattering	SLLS		√	√
Wilhelmy plate			√	√

Table 16 Methods for determining electrokinetic (zeta) potentials.

Name	Acronym	Synonym
Electrokinetic sonic amplitude	ESA	
Electro-osmosis		Electrosmosis, Electric endosmose, Electroendosmosis
Microscopic electrophoresis		Microelectrophoresis, Particle microelectrophoresis
Sedimentation potential		Dorn potential
Spinning cylinder		
Streaming potential		
Ultrasound vibration potential	UVP	Colloid vibration potential (CVP), Colloid vibration current (CVC)

Table 17 Mohs scale of scratch hardness.

Standard material	Non-standard example	Mohs hardness scale value
Talc		1
Gypsum		2
	Fingernail	~2.5
	Gold	~2.5–3
	Penny	~2.5–3
Calcite		3
	Dolomite	3.5
Fluorite		4
	Platinum	~4–4.5
Apatite		5
	Steel knife	~5.5
	Glass	~5.5–6.5
Feldspar		6
	Pyrite	~6.5
	Steel file	~6.5–7.5
Quartz		7
	Garnet	7.5
Topaz		8
	Emerald	8
	Emery cloth	~8.5
Corundum		9
	Ruby	9
	Silicon carbide	~9–10
Diamond		10

Table 18 Porosity ranges.

Classification	Lower size diameter (nm)	Upper size diameter (nm)
Macropores	50	∞
Mesopores	2	50
Micropores	~0	2
Nanopores	~1	100

Source: References [8, 38, 61].

Table 19 Some approximate values of shear rate appropriate to various processes.

Process	Approximate shear rate (s^{-1})	Reference
Very slow stirring	0.01–0.1	[268]
Reservoir flow in oil recovery	1–5	[269]
Mixing	10–100	[184]
Pumping	100–1000	[184]
Coating	10,000	[184]
Oilwell drilling fluid at the bit nozzle	10,000–100,000	[270]

Table 20 Some descriptions appropriate to different yield stresses.

Yield stress (Pa.s)	Description
<10	Easy to pour, like milk
10–20	Thick, pours easily, like thin milkshake
	Use conventional liquid designs
30–40	Thick, hard to pour, forms peaks, can write on surface
	Difficult to make flow under pump suction
40–100	Flows poorly, will cleave to walls under gravity
	Need to push into pump suction
>100	Can build with it, will cleave to top of jar
	Requires positive flow pump

Source: Reference [184].

Table 21 Index of famous names in nanotechnology and colloid and interface science.

Table 21 (*Continued*)

Table 22 Some units and symbols in colloid and interface science.[1]

Quantity	Symbol	Unit name	Abbreviation	Equivalent unit(s)	Older unit(s)
Acceleration	a		$m \cdot s^{-2}$		
Amount of substance	n	mole	mol		
Angle (plane)	θ	radian	rad		$(180/\pi)^\circ$
Angular acceleration	α		$rad \cdot s^{-2}$		
Angular velocity	ω		$rad \cdot s^{-1}$		
Area	A		m^2	10^{-4} ha (hectare)	$10^{20} \times Å^2$
Capacitance (electric)	C	farad	F	$C \cdot V^{-1}$	8.99×10^{11} statfarad
Charge density, volume	ζ, ρ, η		$C \cdot m^{-3}$		3.00×10^3 statcoul $\cdot cm^{-3}$
Compressibility	k		Pa^{-1}	$m \cdot kg^{-1} \cdot s^2$	1.0132×10^5 atm^{-1}
Conductance (electric)	G	siemens	S	$A \cdot V^{-1}$	mho
Conductivity (electric)	γ, σ		$S \cdot m^{-1}$	$A \cdot m^{-1} \cdot V^{-1}$	mho $\cdot m^{-1}$
Current density	J, S		$A \cdot m^{-2}$		
Current density, linear	A, α		$A \cdot m^{-1}$		
Density	ρ		$kg \cdot m^{-3}$		
Diameter	d	metre	m		
Diffusion coefficient	α		$m^2 \cdot s^{-1}$		
Dipole moment, electric	P, p		$C \cdot m$		3.00×10^{11} statcoul $\cdot cm$
Dipole moment, magnetic	j		$N \cdot m^2 A^{-1}$	$Wb \cdot m$	
Distance	d	metre	m		
Electric charge	Q	coulomb	C	$s \cdot A$	3.00×10^9 statcoulomb
Electric current	I	ampere	A		2.998×10^9 statamp
Electric field strength	E, K		$V \cdot m^{-1}$		3.34×10^{-5} statvolt $\cdot cm^{-1}$
Electric flux	ψ, D	coulomb	C	$s \cdot A$	3.00×10^9 statcoulomb
Electric potential	V, φ, ϕ	volt	V	$W \cdot A^{-1}$	3.34×10^{-3} statvolt
Electromotive force	E	volt	V	$W \cdot A^{-1}$	3.34×10^{-3} statvolt
Energy	E, W	joule	J	$N \cdot m$	6.242×10^{18} eV

Table 22 (*Continued*)

Quantity	Symbol	Unit name	Abbreviation	Equivalent unit(s)	Older unit(s)
Enthalpy	H	joule	J		0.2390 calorie
Entropy	S		$J \cdot K^{-1}$		0.2390 calorie $\cdot K^{-1}$
Flow rate, mass	q_m		$kg \cdot s^{-1}$		
Flow rate, volume	q_v		$dm^3 \cdot s^{-1}$	$l \cdot s^{-1}$	
Force	F	newton	N	$m \cdot kg \cdot s^{-2}$	10^5 dyne
Frequency	f, ν	hertz	Hz	s^{-1}	
Heat, quantity	Q	joule	J	$N \cdot m$	0.2390 calorie
Heat capacity	C		$J \cdot K^{-1}$		0.2390 calorie $\cdot K^{-1}$
Heat dissipation coefficient	h, U		$W \cdot m^{-2} K^{-1}$		
Heat flow rate	ϕ, q	watt	W	$J \cdot s^{-1}$	10^7 erg $\cdot s^{-1}$
Heat transfer coefficient	h, U		$W \cdot m^{-2} K^{-1}$		
Illuminance	E	lux	lx	$lm \cdot m^{-2}$	
Impedance, electrical	Z	ohm	Ω		
Inductance	L	henry	H	$Wb \cdot A^{-1}$	
Interfacial tension	γ, Φ		$N \cdot m^{-1}$		10^3 dyne $\cdot cm^{-1}$
Length	l	metre	m		
Light, quantity	Q		$lm \cdot s$	$cd \cdot s \cdot sr$	
Luminance	L		$cd \cdot m^{-2}$		
Luminous flux	Φ	lumen	lm	$cd \cdot sr$	
Luminous intensity	I	candela	cd	$\pi \, cm^2$	π Lambert
Mass	m	kilogram	kg		
Magnetic field strength	H		$A \cdot m^{-1}$		
Magnetic flux	Φ	weber	Wb	$V \cdot s$	
Magnetic flux density	B	tesla	T	$Wb \cdot m^{-2}$	
Magnetic moment	m		$A \cdot m^{-2}$		
Magnetic polarization	B, J	tesla	T	$Wb \cdot m^{-2}$	
Modulus of elasticity	E	pascal	Pa	$m^{-1} \cdot kg \cdot s^{-2}$	9.8692×10^{-6} atm
Molality	m		$mol \cdot kg^{-3}$		
Molar concentration, volume	c		$mol \cdot m^{-3}$		10^{-3} mol $\cdot l^{-1}$
Molar concentration, mass	m		$mol \cdot kg^{-3}$		
Molar mass	M		$kg \cdot mol^{-1}$		
Molar volume	V_m		$m^3 mol^{-1}$		$10^3 l \cdot mol^{-1}$
Momentum	p		$kg \cdot m \cdot s^{-1}$		
Moment of force	M		$N \cdot m$		
Moment of inertia	I		$kg \cdot m^2$		
Permeability, electrical	μ		$H \cdot m^{-1}$		
Permittivity	$\in, \varepsilon$		$F \cdot m^{-1}$		8.99×10^9 statfarad $\cdot cm^{-1}$
Power	P	watt	W	$J \cdot s^{-1}$	10^7 erg $\cdot s^{-1}$
Pressure	P	pascal	Pa	$m^{-1} \cdot kg \cdot s^{-2}$	9.8692×10^{-6} atm
Radiance	L		$W \cdot m^{-2} sr^{-1}$		
Radiant intensity	I		$W \cdot sr^{-1}$		
Radiant flux	ϕ, P	watt	W	$J \cdot s^{-1}$	10^7 erg $\cdot s^{-1}$
Radius	r	metre	m		10^2 cm
Reluctance	R		H^{-1}	$A \cdot Wb^{-1}$	
Resistance (electric)	R	ohm	Ω	$V \cdot A^{-1}$	1.11×10^{-12} statohm

Table 22 *(Continued)*

Quantity	Symbol	Unit name	Abbreviation	Equivalent unit(s)	Older unit(s)
Resistivity (electric)	ζ, ρ		$\Omega \cdot m$	$V \cdot m \cdot A^{-1}$	
Rotational frequency	n		s^{-1}		
Shear modulus	G	pascal	Pa	$m^{-1} \cdot kg \cdot s^{-2}$	9.8692×10^{-6} atm
Specific heat capacity	C		$J \cdot kg^{-1} K^{-1}$		
Stress (normal)	σ	pascal	Pa	$m^{-1} \cdot kg \cdot s^{-2}$	10 dyne $\cdot$ cm^{-2}
Stress (shear)	τ	pascal	Pa	$m^{-1} \cdot kg \cdot s^{-2}$	10 dyne $\cdot$ cm^{-2}
Surface charge density	σ		$C \cdot m^{-2}$		
Surface tension	γ, σ		$N \cdot m^{-1}$		10^3 dyne $\cdot$ cm^{-1}
Temperature	T,t	kelvin	K		$^\circ C + 273$
Thermal conductance	h, U		$W \cdot m^{-2} K^{-1}$		
Thermal conductivity	λ, k		$W \cdot m^{-1} K^{-1}$		
Thermal resistivity			$m \cdot K \cdot W^{-1}$		
Thermal resistance, mechanical	R		$m^2 \cdot K \cdot W^{-1}$		
Thermal resistance, electrical	R		$K \cdot W^{-1}$		
Thickness	δ	metre	m		10^2 cm
Time	t	second	s		
Torque	T		$N \cdot m$		10^7 dyne $\cdot$ cm
Velocity	ν		$m \cdot s^{-1}$		
Viscosity	η, μ		$Pa \cdot s$	$m^{-1} \cdot kg \cdot s^{-1}$	10 P (poise)
Viscosity, kinematic	υ		$m^2 \cdot s^{-1}$		10^4 St (stoke)
Volume	V		m^3	10^3 l (litre)	
Wavelength	λ	metre	m		
Wave number	σ		m^{-1}		
Work	w	joule	J	$N \cdot m$	0.2390 calorie

1) This list is based on the SI system but contains additional units and symbols commonly used in the field. Abbreviations used:

atm = atmosphere
$^\circ$C = degree Celsius
eV = electronvolt
sr = steradian

Appendix 2: References

1 Kerker, M. *J. Colloid Interface Sci.* 1987, **116** (1), 296–299.

2 Asimov, I. *Words of Science and the History Behind Them;* Riverside Press: Cambridge, MA, 1959.

3 Freundlich, H. *Colloid and Capillary Chemistry;* English translation of 3rd. ed., Methuen: London, 1926.

4 Ostwald, W. *An Introduction to Theoretical and Applied Colloid Science;* Wiley: New York, 1917, p. 180.

5 Feynman, R."There's Plenty of Room at the Bottom. An Invitation to Enter a New Field of Physics" presented at: Ann. Meeting, American Physical Society: California Institute of Technology, Dec. 29, 1959.

6 Drexler, E. *Engines of Creation: The Coming Era of Nanotechnology;* Bantam Doubleday Dell, New York, 1986.

7 Crandall, B.C. In *Nanotechnology, Research and Perspectives;* Crandall, B.C.; Lewis, J., eds.; MIT Press, Cambridge, Mass., 1992, p. viii.

8 *Manual of Symbols and Terminology for Physicochemical Quantities and Units;* Appendix II; Prepared by IUPAC Commission on Colloid and Surface Chemistry; Butterworths: London, 1972 and 1978. *See also* the additions in *Pure Appl. Chem.* 1983, 55, 931–941; *Pure Appl. Chem.* 1985, 57, 603–619.

9 Adamson, A. W. *Physical Chemistry of Surfaces;* 5th. ed.; Wiley: New York, 1990.

10 Hiemenz, P.; Rajagopalan, R. *Principles of Colloid and Surface Chemistry,* 3rd. ed.; Dekker, 1997.

11 Shaw, D.J. *Introduction to Colloid and Surface Chemistry,* 3rd. ed.; Butterworths, 1980.

12 Ross, S.; Morrison, I. D. *Colloidal Systems and Interfaces;* Wiley: New York, 1988.

13 Myers, D. *Surfaces, Interfaces, and Colloids;* 2nd. ed.; Wiley-VCH: New York, 1999.

14 Van Olphen, H. and Mysels, K.J., eds. *Physical Chemistry: Enriching Topics from Colloid and Surface Science;* Theorex: La Jolla, CA, 1975.

15 Williams, H. R.; Meyers, C. J. *Oil and Gas Terms;* 6th. ed.; Matthew Bender: New York, 1984.

16 *A Dictionary of Petroleum Terms;* 2nd ed.; Petroleum Extension Service, University of Texas at Austin: Austin, TX, 1979.

17 *The Illustrated Petroleum Reference Dictionary,* 2nd ed.; Langenkamp, R. D., ed.; PennWell Books: Tulsa, OK, 1982.

18 *McGraw-Hill Dictionary of Scientific and Technical Terms;* 3rd. ed.; Parker, S. P., ed.; McGraw-Hill: New York, 1984.

19 Becher, P. *Dictionary of Colloid and Surface Science;* Dekker: New York, 1990.

20 Schramm, L. L., ed. *Emulsions: Fundamentals and Applications in the Petroleum Industry;* American Chemical Society: Washington, DC, 1992.

21 Schramm, L. L., ed. *Foams: Fundamentals and Applications in the Petroleum Industry;* American Chemical Society: Washington, DC, 1994.

22 Schramm, L. L., ed. *Suspensions: Fundamentals and Applications in the Petroleum Industry;* American Chemical Society: Washington, DC, 1996.

23 Schramm, L. L., ed., *Surfactants: Fundamentals and Applications in the Petroleum Industry;* Cambridge University Press: Cambridge, UK, 2000.

Dictionary of Nanotechnology, Colloid and Interface Science. Laurier L. Schramm
Copyright © 2008 WILEY-VCH Verlag GmbH & Co. KGaA, Weinheim
ISBN: 978-3-527-32203-9

24 Becher, P. *Emulsions: Theory and Practice,* 2nd. ed.; Krieger: Malabar, FL, 1977.

25 Whorlow, R. W. *Rheological Techniques;* Ellis Horwood: Chichester, England, 1980.

26 Myers, D. *Surfactant Science and Technology;* VCH: New York, 1988.

27 Rosen, M. J. *Surfactants and Interfacial Phenomena;* Wiley: New York, 1978.

28 Schramm, L.L. *Emulsions, Foams and Suspensions: Fundamentals and Applications;* Wiley-VCH: Weinheim, 2005.

29 *Microemulsions and Emulsions in Foods*; El-Nokaly, M.; Cornell, D., eds.; ACS Symposium Series 448, American Chemical Society: Washington, DC, 1991.

30 *Surface and Colloid Chemistry in Natural Waters and Water Treatment*; Beckett, R., ed.; Plenum Press: New York, 1990.

31 van Olphen, H. *An Introduction to Clay Colloid Chemistry;* 2nd. ed.; Wiley-Interscience: New York, 1977.

32 ASTM Committee E56 on Nanotechnology, Standard Terminology Relating to Nanotechnology, ASTM International: West Conshohocken, PA, 2006.

33 BSI, *Vocabulary – Nanoparticles*, Publicly Available Specification (PAS) 71:2005, British Standards Institution: London, England, 2005.

34 Newton, D.E. *Recent Advances and Issues in Molecular Nanotechnology;* Greenwood Press: Westport, CT, 2002.

35 Bhushan, B. (ed.) *Handbook of Nanotechnology;* Springer-Verlag: Berlin, 2004.

36 Wilson, M.; Kannangara, K.; Smith, G.; Simmons, M.; Raguse, B. *Nanotechnology, Basic Science and Emerging Technologies;* Chapman & Hall/CRC Press: Boca Raton, 2002.

37 Lakhtakia, A., ed. *Nanometer Structures, Theory, Modeling, and Simulation;* ASME Press: N.Y., 2004.

38 Schwarz, J.A.; Contescu, C.I.; Putyera, K. (eds.) *Dekker Encyclopedia of Nanoscience and Nanotechnology,* Volumes 1–4; Dekker: New York, 2004.

39 Asimov, I. *Asimov's Biographical Encyclopedia of Science and Technology;* Doubleday: Garden City, New York, 1964, 192–193.

40 *Biographical Memoirs of Fellows of the Royal Society*; Royal Society (Great Britain): London, 1955.

41 Farber, E. *Nobel prize Winners in Chemistry; Revised ed.,* Abelard-Schuman: London, 1963.

42 Gillispie, C.C. (ed.) *Dictionary of Scientific Biography;* Scribners: New York, 1974.

43 Laidler, K.J. *The World of Physical Chemistry;* Oxford University Press, Oxford, UK, 1993.

44 McMurray, E. J. (ed.) *Notable Twentieth-Century Scientists;* Gale Research Inc.: Detroit, 1995.

45 Babchin, A.J.; Chow, R.S.; Sawatzky, R.P. *Adv. Colloid Interface Sci.* 1989, **30**, 111–151.

46 O'Haver, J.H.; Harwell, J.H. In *Surfactant Adsorption and Surface Solubilization;* American Chemical Society: Washington, 1995, pp. 49–66.

47 Enriquez, L. G.; Flick, G. J. *Dev. Food Sci.* 1989, **19**, 235–334.

48 Anderson, W. G. *J. Petrol. Technol.* 1986, **38** (12), 1246–1262.

49 Ramsey, M. *Glossary Of Lubrication, Filtration and Oil Analysis Terms;* World Wide Web page, http://www.diagnetics.com/glossary.html; Diagnetics, 1996.

50 Somorjai, G.A. *Introduction to Surface Chemistry and Catalysis,* Wiley-Interscience: New York, 1994.

51 Adam, N.K. *Physical Chemistry;* Oxford University Press: Oxford, pp. 577–635, 1956.

52 Isenberg, C. *The Science of Soap Films and Soap Bubbles;* Tieto Ltd.: Clevedon, United Kingdom, 1978, p 101.

53 Carrrière, G. *Dictionary of Surface Active Agents, Cosmetics and Toiletries;* Elsevier: Amsterdam, 1978.

54 Antonow, G. *J. Chim. Phys.* 1907, **5**, 372.

55 *Glossary of Soil Science Terms*; Soil Science Society of America: Madison, WI, 1987.

56 *Glossary of Terms in Soil Science*; Research Branch, Agriculture Canada: Ottawa, Ontario; revised 1976.

57 ASTM Committee E42 on Surface Analysis, *Standard Terminology Relating to Surface Analysis;* ASTM International: West Conshohocken, PA, 2003.

58 Gillispie, C.C. (ed.) *Dictionary of Scientific Biography,* Volume I; Scribners: New York, 1970, pp. 430–431.

59 *National Academy of Sciences Biographical Memoirs*, Vol 65; National Academy Press: Washington, DC, 1994, pp. 3–39.

60 Barrett, E.P., Joyner, L.G., Halenda, P.P. *J. Amer. Chem. Soc.* 1951, **73**, 373–380.

61 Lowell, S., Shields, J.E. *Powder Surface Area and Porosity;* Chapman and Hall: London, 1991.

62 Saliterman, S.S. *BioMEMS and Medical Microdevices;* Wiley-Interscience: New York, 2006.

63 Hartmann, U., Mende, H.H. *Z. Phys. B Condensed Matter*, 1985, **61**, 29–32.

64 Martinez, A. R. In *The Future of Heavy Crude and Tar Sands;* Meyer, R. F.; Wynn, J. C.; Olson, J. C., eds.; UNITAR: New York, 1982; pp. ixvii–ixviii.

65 Danyluk, M., Galbraith, B., Omana, R. In *The Future of Heavy Crude and Tar Sands;* Meyer, R. F.; Wynn, J. C.; Olson, J. C., eds.; UNITAR: New York, 1982; pp. 3–6.

66 Khayan, M. In *The Future of Heavy Crude and Tar Sands;* Meyer, R. F.; Wynn, J. C.; Olson, J. C., eds.; UNITAR: New York, 1982; pp. 7–11.

67 McMurray, E.J. (ed.) *Notable Twentieth-Century Scientists,* Volume 1; Gale Research Inc.: Detroit, 1995, pp. 197–199.

68 Kass-Simon, G.; Farnes, P. (eds.) *Women of Science;* Indiana University Press: Bloomington, 1990, pp. 198–199.

69 Binnington, R.J.; Boger, D.V. *J. Rheol.* 1985, **29** (6), 887–904.

70 *McGraw-Hill Modern Scientists and Engineers,* Volume 1; McGraw-Hill: N.Y., 1980, pp. 118–119.

71 Meunier, *J. Coll. Surf. A*, 2000, **171**, 33–40.

72 Asimov, I. *Asimov's Biographical Encyclopedia of Science and Technology;* Doubleday: Garden City, N.Y., 1964, pp. 192–193.

73 Taylor, H.S. In *Solid Surfaces and the Gas-Solid Interface;* American Chemical Society: Washington, 1961, pp. 2–4.

74 Kerker, M.; Zettlemoyer, A.C., *J. Colloid Interface Sci.*, 1986, **114**, 300–302.

75 Wadle, A.; Tesmann, H.; Leonard, M.; Förster, T. In *Surfactants in Cosmetics,* 2nd. ed.; Rieger, M.M.; Rhein, L.D., eds.; Dekker: New York, 1997, pp. 207–224.

76 Britz, D.A.; Khlobystov, A.N.; Porfyrakis, K.; Ardavan, A.; Briggs, G.A.D. Chem. Commun., **2005**, *advance article posted online Nov. 18, 2004,* 5 pp.

77 Carreau, P.J. *Trans. Soc. Rheol.* 1972, **16**, 99–127.

78 Gillispie, C.C. (ed.) *Dictionary of Scientific Biography,* Volume 3; Scribners: New York, 1971, pp. 197.

79 *Biographical Memoirs of Fellows of the Royal Society,* 1958, 4, 35–44.

80 Kukla, R. *J. Chem. Eng. Progr.* 1991, **87**, 23–35.

81 Stewart, S.A. *A Glossary of Urethane Industry Foams;* Martin Sweets Co., Inc.: Louisville, KY, 1971.

82 Dinsmore, A.D.; Hsu, M.F.; Nikolaides, M.G.; Marquez, M.; Bausch, A.R.; Weitz, D.A. *Science,* 2002, **298**, 1006–1009.

83 "Wallace Henry Coulter," http://www.whcf.org/WHCF_WallaceHCoulter.htm, Wallace Henry Coulter Foundation, Maimi, FL, accessed July 23, 2005.

84 Mukerjee, P. and Mysels, K.J. *Critical Micelle Concentrations of Aqueous Surfactant Systems,* Nat. Stand. Ref. Data Ser. NSRDS-NBS 36, National Bureau Standards: Washington, D.C., 1971.

85 van Os, N.M., Haak, J.R., and Rupert, L.A.M. *Physico-Chemical Properties of Selected Anionic, Cationic and Nonionic Surfactants;* Elsevier: Amsterdam, 1993.

86 Gillispie, C.C. (ed.) *Dictionary of Scientific Biography,* Volume III; Scribners: New York, 1971, pp. 617–621.

87 Farber, E. *Nobel prize Winners in Chemistry;* Revised ed., Abelard-Schuman: London, 1963, pp. 147–151.

88 Kitchener, J.A. In *Surface Forces;* Derjaguin, B.V.; Churaev, N.V.; Muller, V.M., eds., English translation, Consultants Bureau: New York, 1987, pp. ix–xiv.

89 Churaev, N.V.; Vinogradova, O.I. *J. Colloid Interface Sci.*, 1994, **168**, 273–274.

90 Mirkin, C.A. *ACS Nano* 2007, **1** (2), 79–83.

91 Drexler, K.E. *Proc. Natl. Acad. Sci. (USA)*, 1981, **78** (9), 5275–5278.

92 Drexler, K.E. *Engines of Creation*, Fourth Estate; London, 1990.

93 Bikerman, J.J. *Foams;* Springer-Verlag: New York, NY, 1973.

94 Kerker, M. *J. Colloid Interface Sci.* 1989, **129** (1), 291–295.

95 Asimov, I. *Asimov's Biographical Encyclopedia of Science and Technology;* Doubleday: Garden City, New York, 1964, pp. 483–487.

96 Hunter, R.J. *Coll. Surf. A* 1998, **141**, 37–65.

97 Donini, J.C.; Angle, C.W.; Hassan, T.A.; Kasperski, K.L.; Kan, J.; Kar, K.L.; Thind, S.S. In *Emerging Separation Technologies for Metals* *and Fuels;* Lakshmanan, V.I.; Bautista, R.G.; Somasundaran, P., eds.; The Minerals, Metals & Materials Society, 1993, pp. 409–424.

98 Donald, A.M.; Litching, S.; Meredith, P.; He, C. In *Modern Aspects of Colloidal Dispersions;* Ottewill, R.H.; Rennie, A.R., eds.; Kluwer Publishers: Dordrecht, 1998, pp. 41–50.

99 Gast, A.P.; Zukoski, C.F. *Adv. Colloid Interface Sci.* 1989, **30**, 153–202.

100 *McGraw-Hill Modern Scientists and Engineers,* Volume 1; McGraw-Hill: N.Y., 1980, pp. 337–338.

101 *National Academy of Sciences Biographical Memoirs,* Vol. 67; National Academy Press: Washington, DC, 1995, pp. 119–219.

102 Tadros, Th.F. In *Handbook of Applied Surface and Colloid Chemistry;* Holmberg, K. (ed.), Vol. 1; Wiley: New York, 2001, pp. 73–83.

103 Dalgleish, D.G. In *Emulsions and Emulsion Stability;* J. Sjoblom (ed.), Marcel Dekker: New York, 1996, pp. 287–325.

104 Cash, L.; Cayias, J. L.; Fournier, G.; Macallister, D.; Schares, T.; Schechter, R. S.; Wade, W. H. *J. Colloid Interface Sci.* 1977, **59**, 39–44.

105 Cayias, J. L.; Schechter, R. S.; Wade, W. H. *Soc. Petrol. Eng. J.* 1976, **16**, 351–357.

106 Kerker, M. *J. Colloid Interface Sci.* 1986, **112** (1), 302–305.

107 Weiser, H.B. *Inorganic Colloid Chemistry,* Volume 1; Wiley: New York, 1933.

108 Elias, F.; Bacri, J-C.; Flament, C.; Janiaud, E.; Talbot, D.; Drenckhan, W.; Hutzler, S.; Weaire, D. *Coll. Surf. A,* 2005, **263**, 65–75.

109 Feynman, R.P. *Engineering and Science (Calif. Inst. Technolol.),* **1960**, Feb.; reprinted in Hey, A. J. G., ed., *Feynman and Computation;* Perseus Books: Reading, MA, 1998; also at http://www.zyvex.com/nanotech/feynman.html.

110 *McGraw-Hill Modern Scientists and Engineers,* Volume 1; McGraw-Hill: New York, 1980, pp. 382–383.

111 Dickinson, E. *An Introduction to Food Colloids;* Oxford University Press: Oxford, 1992.

112 Gillispie, C.C. (ed.) *Dictionary of Scientific Biography,* Volume XV, Supplement 1; Scribners: New York, 1978, pp. 159–160.

113 Freundlich, H. *Colloid and Capillary Chemistry;* Methuen: London, 1926.

114 Healy, T.W.; Han, K.; King, P. *Int. J. Miner. Process.* 2003, **72**, 1–2.

115 Fuerstenau, D.W. *Froth Flotation;* Amer. Inst. Mining, Mech. & Petrol Eng. (AIME): New York, 1962.

116 Lipatov, Y.S. *Colloid Chemistry of Polymers;* Elsevier, 1988.

117 Kruyt, H.R. In *Colloid Science;* Kruyt, H.R., ed., Volume 1; Elsevier, pp. 1–57, 1952.

118 Everett, D.H., *Basic Principles of Colloid Science;* Royal Society of Chemistry, 1988.

119 Almdal, K.; Dyre, J.; Hvidt, S.; Kramer, O. *Polymer Gels Networks* 1993, **1**, 5–17.

120 Wheeler, L.P. *Josiah Willard Gibbs;* Archon Books, 1970.

121 Seeger, R.J. J. *Willard Gibbs;* Pergamon Press: Oxford, England, 1974.

122 Gillispie, C.C. (ed.) *Dictionary of Scientific Biography,* Volume 5; Scribners: New York, 1972, pp. 483–484.

123 Kerker, M. *J. Colloid Interface Sci.* 1987, **116** (1), 296–299.

124 Asimov, I. *Words of Science and the History Behind Them;* Riverside Press: Cambridge, Mass., 1959, p. 58.

125 O'Lenick, A.J. *J. Surf. Deter.* 2001, **4**, 311–315.

126 Frisch, H.L.; Simha, R. In *Rheology Theory and Applications;* Eirich, F.R., ed., Volume 1; Academic Press: New York, 1956; pp. 525–613.

127 Mysels, K.J.; Scholten, P.C. *Langmuir* 1991, **7**, 209–211.

128 Gillispie, C.C. (ed.) *Dictionary of Scientific Biography,* Volume VI; Scribners: New York, 1974, pp. 117–119.

129 *National Academy of Sciences Biographical Memoirs,* Volume 47; National Academy Press: Washington, DC, 1975, pp. 49–81.

130 Harwell, J. H.; Hoskins, J. C.; Schechter, R. S.; Wade, W. H. *Langmuir* 1985, **1** (2), 251–262.

131 Mulligan, J.F. (ed.) *Heinrich Rudolf Hertz (1857–1894);* Garland Publishing: New York (1994).

132 *Biographical Memoirs of Fellows of the Royal Society,* 1982, 28, 153–162.

133 Schulz, D.N.; Kaladas, J.J.; Maurer, J.J.; Bock, J.; Pace, S.J.; Schulz, W.W. *Polymer* 1987, **28**, 2110–2115.

134 Fuerstenau, D. W.; Williams, M. C. *Colloids Surf.* 1987, **22**, 87–91.

135 Iijima, S. *Nature,* 1991, **354**, 56-58.

136 Ajayan, P.M.; Iijima, S. *Nature,* 1992, **358**, 23–23.

137 Servos, J.W. *Physical Chemistry from Ostwald to Pauling: The Making of a Science in America;* Princeton University Press: Princeton, NJ, 1990, **402**, pp.

138 Zeece, M. *Trends Food Sci. Technol.* 1992, **3**, 6–10.

139 Jacobs, M. *Chem. & Eng. News*, 1996, (*January* 22), 5.

140 Krafft, F.; Wiglow, H. *Ber. Dtsch. Chem. Ges.* 1895, **28**, 2566–2573.

141 Kreiger, I.M.; Dougherty, T.J. *Trans. Soc. Rheol.*, 1959, **3**, 137.

142 Schramm, L. L.; Novosad, J. J. *Colloids Surf.* 1990, **46**, 21–43.

143 Ter Haar, D. *Collected Papers of L.D. Landau;* Gordon and Breach: New York, 1965.

144 Khalatnikov, I.M. (ed.) *Landau the Physicist and the Man;* Pergamon Press: Oxford, England, 1989.

145 Kerker, M. *J. Colloid Interface Sci.* 1989, **133** (1), 290–292.

146 Greene, J.E. (ed.) *100 Great Scientists;* Washington Square Press: New York, 1964.

147 Carvill, J. *Famous Names in Engineering;* Butterworths: London, 1981.

148 Gunstone, F.D. In *Lipid Technologies and Applications;* Gunstone, F.D.; Padley, F.B., eds.; Marcel Dekker: New York, 1997, pp. 1–12.

149 Beck, J.S., Vartuli, J.C., Roth, W.J., Leonowicz, M.E., Kresge, C.T., Schmitt, K.D., Chu, C.T-W., Olson, D.H., Sheppard, E.W., McCullen, S.B., Higgins, J.B., Schlenker, J.L. *J. Amer. Chem. Soc.* 1992, **114**, 10834–10843.

150 Rasa, M.; Kuipers, B.W.M.; Philipse, A.P. *J. Colloid Interface Sci.* 2002, **250**, 303–315.

151 Schramm, L. L.; Clark, B. W. *Colloids Surf.* 1983, **7**, 135–146.

152 Adamson, A.W. and Gast, A.P. *Physical Chemistry of Surfaces,* 6th. ed.; Wiley-Interscience: New York, 1997.

153 Gunstone, F.D. and Norris, F.A. *Lipids in Foods. Chemistry, Biochemistry and Technology;* Pergamon Press: New York, 1983.

154 *National Academy of Sciences Biographical Memoirs,* Volume 68; National Academy Press: Washington, DC, 1995, pp. 195–208.

155 Porter, R. (ed.) *The Hutchinson Dictionary of Scientific Biography;* Helicon: Oxford, GB, 1994, pp. 453–454.

156 Graham, N.B.; Cameron, A. *Pure & Appl. Chem.,* 1998, **70** (6), 1271–1275.

157 Reyes, D.R.; Iossifidis, D.; Auroux, P-A.; Manz, A. *Anal. Chem.* 2002, **74**, 2623–2636.

158 Gillispie, C.C. (ed.) *Dictionary of Scientific Biography,* Volume IX; Scribners: New York, 1974, pp. 376–377.

159 Mukerjee, P. and Hofmann, A.F. *J. Colloid Interface Sci.* 2000, **224**, 1–3.

160 Ksomasundaran, P.; Goddard, E.D.; Pethica, B.P. *Colloids and Surfaces A,* 2001, **186**, 3–5.

161 Poncharal, P.; Wang, Z.L.; Ugarte, D.; de Heer, W.A. *Science* 1999, **283** (5407), 1513–1516.

162 Ozin, G.A.; Arsenault, A.C. *Nanochemistry, A Chemical Approach to Nanomaterials;* Royal Society of Chemistry: Cambridge, UK, 2005.

163 Hunt, G. In *Nanotechnology, Risk, Ethics and Law,* Hunt, G.; Mehta, M.D. (eds.); Earthscan: London, UK, 2006, pp. 43–56.

164 Bonini, M.; Lenz, S.; Giorgi, R.; Baglioni, P. *Langmuir,* 2007, **23** (17), 8681–8685.

165 Fennimore, A.M.; Yuzvinsky, T.D.; Han, W.; Fuhrer, M.S.; Cumings, J.; Zettl, A. *Nature* 2003, **424**, 408–410.

166 Montemagno, C.; Bachand, G. *Nanotechnology* 1999, **10**, 225–231.

167 Ahuja, A.; Taylor, J.A.; Lifton, V.; Sidorenko, A.A.; Salamon, T.R.; Lobaton, E.J.; Kolodner, P.; Krupenkin, T.N. *Langmuir* 2008, **24** (1), 9–14.

168 Black, C.T. *ACS Nano* 2007, **1** (3), 147–150.

169 Novoselov, K.S.; Geim, A.K.; Morozov, S.V.; Jiang, D.; Zhang, Y.; Dubonos, S.V.; Grigorieva, I.V.; Firsov, A.A. *Science,* 2004, **306**, 666–669.

170 Vaidya, R.; López, G.; López, J.A. In *Kirk-Othmer Encyclopedia of Chemical Technology,* Online edition, posted December 4, 2000; Wiley: New York, 1998.

171 Kolesnikov, A.I.; Zanotti, J.; Loong, Chun.; Thiyagarajan, P.; Moravsky, A.P.; Loutfy, R.O.; Burnham, C.J. *Phys. Rev. Let.* 2004, **93**, article 035503.

172 Srivastava, A.; Srivastava, O.N.; Talapatra, S.; Vajtai, R.; Ajayan, P.M. *Nature Mat.,* 2004, **3**, 610–614.

173 Hinds, B.J.; Chopra, N.; Rantell, T.; Andrews, R.; Gavalas, V.; Bachas, L.G. *Science,* 2004, **303**, 62–65.

174 Wang, Z.L. (ed.) *Nanowires and Nanobelts, Materials, Properties, and Devices,* Vols. 1 and 2; Kluwer Academic: Boston, 2003.

175 Wu, Y.; Xiang, J.; Yang, C.; Lu, W.; Lieber, C.M. *Nature*, 2004, **430**, 61–65.

176 Whang, D.; Jin, S.; Wu, Y.; Lieber, C.M. *Nano Lett.*, 2003, **3**, 1255–1259.

177 Nelson, R. C. *Chem. Eng. Prog.* 1989, *March*, 50–57.

178 McMurray, E.J. (ed.) *Notable Twentieth-Century Scientists*, Volume 3; Gale Research Inc.: Detroit, 1995, pp. 1461–1464.

179 Becher, P. *J. Colloid Interface Sci.* 1990, **140** (1), 300–301.

180 Laskowski, J.S. In *Colloid Chemistry in Mineral Processing*; Laskowski, J.S.; Ralston, J. (eds.); Elsevier, Amsterdam, 1992, pp. 361–394.

181 Panizza, P.; Roux, D.; Vuillaume, V.; Lu, C.-Y.D.; Cates, M.E. *Langmuir*, 1996, **12**, 248–252.

182 National Academy of Sciences Biographical Memoirs, Volume 60; National Academy Press: Washington, DC, 1991, pp. 183–232.

183 Shen, Y.R. *Nature* 1989, **337**, 519–525.

184 Corn, R.M.; Higgins, D.A. *Chem. Rev.* 1994, **94**, 107–125.

185 Garbow, N.; Evers, M.; Palberg, T. *Coll. Surf. A*, 2001, **195**, 227–241.

186 Grier, D.G. *Curr. Opinion Coll. Interface Sci.* 1997, **2**, 264–270.

187 Mackie, A.R.; Gunning, A.P.; Wilde, P.J.; Morris, V.J. *J. Coll. Interface Sci.* 1999, **210**, 157–166.

188 Gillispie, C.C. (ed.) *Dictionary of Scientific Biography*, Volume X; Scribners: New York, 1974, pp. 251–252.

189 Gillispie, C.C. (ed.) *Dictionary of Scientific Biography*, Volume XV; Suppl. I, Scribners: New York, 1978, pp. 455–469.

190 Servos, J.W. *Physical Chemistry from Ostwald to Pauling: The Making of a Science in America*; Princeton University Press: Princeton, NJ, 402 pp. 1990.

191 Kerker, M. *J. Colloid Interface Sci.* 1987, **115** (1), 291–294.

192 Gillispie, C.C. (ed.) *Dictionary of Scientific Biography*, Volume 10; Scribners: New York, 1974, pp. 524–526.

193 Beisswanger, G. *Chem. Unserer Zeit* 1991, **25** (2), 96–101.

194 Kass-Simon, G.; Farnes, P. (eds.) *Women of Science*; Indiana University Press: Bloomington, 1990, pp. 309–310.

195 Derrick, M.E. *J. Chem. ed.* 1982, **59** (12), 1030–1031.

196 Halverson, F.; Panzer, H.P. In *Kirk-Othmer Encyclopedia of Chemical Technology*; 3rd. ed.; Wiley: New York, 1980, **10**, 489–523.

197 van den Hul, H.J.; Vanderhoff, J.W. In *Polymer Colloids*; Fitch, R.M., ed.; Plenum: New York, 1971, pp. 1–27.

198 Kraynik, A. *Coll. Surf. A*, 2005, **263**, 1–2.

199 Doraiswamy, D. *Rheology Bull.* 2002, **71** (1), 9pp.

200 Freundlich, H.; Juliusberger, F. *Trans. Faraday Soc.*, 1935, **31**, 920.

201 Nguyen, A.V. *Int. J. Miner. Process.* 1999, **56**, viii–x.

202 Stöckelhuber, K.W. *Adv. Coll. Interface Sci.* 2005, **114/115**, 1–2.

203 Verwey, E.J.W.; Overbeek, J.T.G. *Theory of the Stability of Lyophobic Colloids*; Elsevier, 1948.

204 Freundlich, H. *Colloid and Capillary Chemistry*; Methuen, 1926.

205 Horn, R. In *Science, Technology, and Competitiveness*; Hixenbaugh, G.W. (ed.), National Institute of Standards and Technology: Washington, 1996, pp. 96–104.

206 Benee, L.S.; Snowden, M.J.; Chowdhry, B.Z. In *Encyclopedia of Polymer Science and Technology*; Online edition, posted March 15, 2002; Wiley: New York, 2002.

207 Snowden, M. *Science Spectra* 1996, **6**, 32–35.

208 Snowden, M.; Murray, M.J.; Chowdry, B.Z. *Chem. & Ind.* 1996, 15 July, 531–534.

209 Kerker, M. *J. Colloid Interface Sci.* 1988, **124** (2), 697–699.

210 Lea, J.; Nickens, H.V.; Wells, M. *Gas Well Deliquification*; Gulf Professional Publishing: Burlington, MA, 2003.

211 Evans, D.F.; Wennerstrom, H. *The Colloidal Domain*; VCH, 1994.

212 Brinker, C.J.; Scherer, G.W. In *Sol-Gel Science*; Academic Press: New York, 1990.

213 Spelt, J. K. *Colloids Surf.* 1990, **43**, 389–411.

214 Somasundaran, P.; Krishnakumar, S.; Kunjappu, J.T. In *Surfactant Adsorption and Surface Solubilization*; American Chemical Society: Washington, 1995; pp. 104–137.

215 Kosmulski, M.; Matijević, E. *J. Colloid Interface Sci.* 1992, **150** (1), 291–294.

216 Ananthapadmanabhan, K.P.; Moudgil, B.M. *J. Coll. Interface Sci.* 2002, **256**, 1–2.

217 Capes, C. E.; McIlhinney, A. E.; Sirianni, A. F. In *Agglomeration 77;* American Institute of Mining, Metallurgical, and Petroleum Engineers: New York, 1977, pp. 910–930.

218 Flack, E. In *Lipid Technologies and Applications;* Gunstone, F.D.; Padley, F.B. (eds.); Dekker, New York, 1997, pp. 305–327.

219 Ross, S.; Becher, P. *J. Colloid Interface Sci.* 1992, **149**, 575–579.

220 McMurray, E.J. (ed.) *Notable Twentieth-Century Scientists,* Volume 4; Gale Research Inc.: Detroit, 1995, pp. 1907–1910.

221 Mysels, K.J. *Langmuir* 1989, **5**, 539.

222 Kerker, M. *J. Colloid Interface Sci.* 1986, **113** (2), 589–593.

223 Strutt, R.J. *Life of John William Strutt Third Baron Rayleigh,* 2nd. ed.; University of Wisconsin Press: Madison, 1968.

224 Luckham, P.F.; de Costello, B.A. *Adv. Colloid Interface Sci.* 1993, **44**, 183–240.

225 Horn, R. In *Science, Technology, and Competitiveness;* Special Publication 837; National Institute of Standards and Technology: Washington, DC, 1996, pp. 96–104.

226 Israelachvili, J. *Intermolecular and Surface Forces,* 2nd. ed.; Academic Press: San Diego, 1991.

227 Dorshow, R.B.; Swofford, R.L. *Colloids Surf.* 1990, **43**, 133–149.

228 Krog, N.; Barfod, N.M.; Sanchez, R.M. *J. Disp. Sci. Technol.* 1989, **10**, 483–504.

229 Stevens, C.E. In *Kirk-Othmer Encyclopedia of Chemical Technology,* 2nd. ed., Volume 19; Wiley: New York, 1969, pp. 507–593.

230 Salager, J.L.; Marquez, N.; Graciaa, A.; Lachaise, J. *Langmuir* 2000, **16**, 5534–5539.

231 Kerker, M. *J. Colloid Interface Sci.* 1986, **111** (1), 295–297.

232 Ranby, B. *Macromol. Symp.* 1995, **98**, 1227–1245.

233 Benedek, G.B.; Kardar, M.; Litster, J.D. *Physics Today Online* 2001, February, 3 pp.

234 Taniguchi, N. *On the Basic Concept of 'Nano-Technology';* Proc. Intl. Conf. Prod. Eng. Tokyo, Part II, Japan Society of Precision Engineering, 1974.

235 Edwards, S.A. *The Nanotech Pioneers, Where Are They Taking Us?* Wiley-VCH: Weinheim, Germany, 2006.

236 McMurray, E.J. (ed.) *Notable Twentieth-Century Scientists,* Volume 4; Gale Research Inc.: Detroit, 1995, pp. 1986–1989.

237 Blumberg, S.A. and Panos, L.G. (eds.) *Edward Teller Giant of the Golden Age of Physics;* Scribners: New York, 1990.

238 Balasubramaniam, R.; Lacy, C.E.; Woniak, G.; Subramanian, R.S., *Phys, Fluids* 1996, **8**, 872–880.

239 Freundlich, H.; Bircumshaw, L.L., *Kolloid Z.,* 1926, **40**, 19.

240 Thompson, S.P. *The Life of William Thomson, Baron Kelvin of Largs,* 2 Vols.; Macmillan: London, 1910.

241 O'Lenick, A.J.; Parkinson, J.K. *Cosmetics & Toiletries,* 1997, **112** (11), 59–64.

242 Figueredo, R.C.R.; Sabadini, E. *Coll. Surf. A,* 2003, **215**, 77–86.

243 Figueredo, R.C.R.; Sabadini, E. *Coll. Surf. A,* 2003, **215**, 77–86.

244 Gregory, J. *Particles in Water. Properties and Processes;* CRC Press: Boca Raton, FL, 2006.

245 Turner, R.G.; Hubbard, M.G.; Dukler, A.E. *J. Petrol. Technol.,* 1969 (*Nov.*), 1475–1482.

246 Kerker, M. *J. Colloid Interface Sci.* 1987, **119** (2), 602–604.

247 Eve, A.S.; Creasey, C.H. *Life and Work of John Tyndall;* Macmillan: London, 1945.

248 Gillispie, C.C. (ed.) *Dictionary of Scientific Biography,* Volume XIV; Scribners: New York, 1974, pp. 109–111.

249 Drexler, K.E. In *Nanotechnology, Research and Perspectives;* Crandall, B.C.; Lewis, J. (eds.); MIT Press, Cambridge, Mass., 1992, pp. 325–346.

250 Farber, E. *Nobel prize Winners in Chemistry;* Revised ed.; Abelard-Schuman: London, 1963, pp. 1–4.

251 Drost-Hansen, W.; Singleton, J.L. In *Chemistry of the Living Cell;* JAI Press, 1992; pp. 157–180.

252 Etzler, F.M.; Fagundus, D.M. *J. Colloid Interface Sci.* 1987, **115** (2), 513–519.

253 Shearer, B.F.; Shearer, B.S. *Notable Women in the Physical Sciences: A Biographical Dictionary;* Greenwood Press: Westport, CT, 1997, pp. 401–404.

254 Proffitt, P. (ed.) *Notable Women Scientists;* Gale Group: Detroit, 1999, pp. 596–597.

255 Li, C.B.; Friedman, S.P. *Coll. Surf. A,* 2003, **222**, 133–140.

256 Gillispie, C.C. (ed.) *Dictionary of Scientific Biography,* Volume 14; Scribners: New York, 1976, pp. 359–360.

257 Wilson, C.T.R. *Phil. Trans. Roy. Soc. (London) A* 1897, **189**, 265; 1900, **193**, 289.

258 Winsor, P. A. *Solvent Properties of Amphiphilic Compounds;* Butterworths: London, 1954.

259 Warneck, P. *Chemistry of the Natural Atmosphere;* Academic: San Diego, CA, 1988.

260 Rogers, R. R.; Yau, M. K. *A Short Course in Cloud Physics,* 3rd. ed.; Pergamon: Oxford, England, 1989.

261 Hidy, G.M.; Brock, J.R. *The Dynamics of Aerocolloidal Systems;* Pergamon Press: Oxford, 1970.

262 Woodruff, D. P.; Delchar, T. A. *Modern Techniques of Surface Science;* Cambridge University Press: New York, 1986.

263 Somorjai, G.A., *Introduction to Surface Chemistry and Catalysis*; Wiley-Interscience: New York, 1994.

264 Scholle, P. A. *Constituents, Textures, Cements, and Porosities of Sandstones and Associated Rocks; Memoir 28;* American Assocciation of Petroleum Geologists: Tulsa, OK, 1979.

265 Blatt, H.; Middleton, G.; Murray, R. *Origin of Sedimentary Rocks;* 2nd. ed.; Prentice-Hall: Englewood Cliffs, NJ, 1980.

266 La Mesa, C. *J. Phys. Chem.* 1990, **94**, 323–326.

267 Stasiuk, E.N.; Schramm, L.L. *J. Colloid Interface Sci.,* 1996, **178**, 324–333.

268 Barnes, H.A.; Holbrook, S.A. In *Processing of Solid-Liquid Suspensions,* Ayazi Shamlou, P., ed.; Butterworth Heinemann: Oxford, UK, 1993, pp. 222–245.

269 Lake, L.W. *Enhanced Oil Recovery,* Prentice Hall: Englewood Cliffs, NJ, 1989, pp. 47–48, 322.

270 Clark, R.K.; Nahm, J.J. In *Kirk-Othmer Encyclopedia of Chemical Technology,* 3rd ed., Volume 17; Wiley: New York, 1982, pp. 143–167.